Photographs Across Time: Studies in Urban Landscapes

Authored by

Mary J. Thornbush

School of Geography
Earth and Environmental Sciences
University of Birmingham
UK

&

Sylvia E. Thornbush

School of History
Classics and Archaeology
University of Edinburgh
UK

PERMISSION/DISCLAIMER

CONTENTS

FOREWORD

The use of repeat photography in geomorphology has become a well-established technique for illustrating landscape changes. Its use has, however, largely been restricted to relatively 'natural' landscapes in large parks in the American west such as Grand Canyon, Yellowstone, and Glacier National Parks. In this volume, authors Mary J. Thornbush and Sylvia E. Thornbush apply the techniques of what they refer to as 'photo geomorphology' to the very different landscapes of urban portions of England and Scotland. In this well-illustrated and fascinating volume, the authors use the techniques of photogeomorphology to document changes to historic structures in Oxford, England, and to urban churchyards in England and Scotland *via* detailed photo-documentation of changes to headstones/ gravestones. Bringing to bear their expertise in applied geomorphology (M. J. Thornbush) and urban archaeology (S. E. Thornbush), the authors illustrate how 'archaeogeomorphology' provides a more thorough understanding of the physical and cultural landscape than traditional observations. They do so through the use of quantitative studies of urban heritage sites as well as qualitative studies of the historical archaeology of urban churchyards. Together, the techniques and expertise on display in this book make a compelling argument for the desirability of broader applications of photogeomorphology in other locations around the world. I congratulate the authors on the spirit of enquiry on display in this book, and strongly commend it to readers across the fields of geomorphology, physical geography, landscape and heritage studies, and urban archaeology.

David R. Butler
Texas State University System Regents' Professor of Geography
University Distinguished Professor of Geography
Honorary Professor of International Studies
Department of Geography
Texas State University
San Marcos
Texas 78666
USA

PREFACE

This eBook aims to apply a photogeomorphological approach to landscape studies for central Oxford and urban churchyards in England and Scotland. It embraces an integrated multidisciplinary approach in complementary fields within applied geomorphology (photogeomorphology and landscape studies) and historical archaeology in urban settings. Its scope comprises the use of digital photographs to capture and convey cross-temporally landscape change in urban areas in the UK. A variety of topics are introduced initially and developed on a chapter-by-chapter basis by the authors. Part I is written by the first-listed author from an applied geomorphology perspective, and Part II is written by the second-listed author as a historical archaeologist. The former author has been working on Oxford studies since 2002 (and published since 2004) and the latter author began an examination of urban British churchyards predominantly towards the East Coast since 2007. Each author brings a different disciplinary perspective to this eBook; it is hoped that this will benefit an integrated presentation of physical and cultural landscapes that encompasses traditional geomorphology, as well as more applied geomorphology that encompasses 'archaeogeomorphology' within an environmental geomorphology framework, with inputs from historical archaeology and landscape studies.

CONFLICT OF INTEREST

The authors confirm that this chapter contents have no conflict of interest.

Mary J. Thornbush
School of Geography
Earth and Environmental Sciences
University of Birmingham
UK
E-mail: m.thornbush@bham.ac.uk

&

Sylvia E. Thornbush
School of History
Classics and Archaeology
University of Edinburgh
UK

List of Contributors

Mary J. Thornbush

School of Geography, Earth and Environmental Sciences, University of Birmingham, Edgbaston, Birmingham, West Midlands, B15 2TT, England, UK

Dr Thornbush has published many studies on rock weathering and the conservation of Oxford's historical buildings. She is a trained geomorphologist (since 1999), who studied at the University of Toronto, Canada, and completed a doctoral degree in the School of Geography and the Environment at the University of Oxford, UK. In 2006-2007, she was a postdoctoral research assistant on a project funded by English Heritage in the Oxford Rock Breakdown Laboratory. She remained affiliated as a senior research associate of the Oxford University Centre of the Environment until 2008. She has since returned to reside in central Oxford and continues to publish local studies, including photogeomorphological research. Her current research interests include applied geomorphology and landscape studies. Dr Thornbush is currently an honorary research fellow at the School of Geography, Earth and Environmental Sciences at the University of Birmingham, UK.

Sylvia E. Thornbush

School of History, Classics and Archaeology, University of Edinburgh, William Robertson Wing, Old Medical School, Teviot Place, Edinburgh, EH8 9AG, Scotland, UK

S. E. Thornbush has been interested in headstone studies since 1999, when she began in-depth investigations into the seriation of headstone styles in North America and now in Scotland and England. She is interested in the different societal influences on headstone designs from the 17th century to the late 19th century. Ms Thornbush has presented at several conferences on her seriation research and has published in this area. She is due to complete her PhD at the University of Edinburgh in Scotland, UK in 2015, and will then continue to research at the postdoctoral level. Her research interests include symbology, historical architecture, and feminist archaeology.

CHAPTER 1

INTRODUCTION

Abstract: This book examines the use of photography across two different but complementary disciplines. The first part focuses on the use of photography in landscape studies, revisiting several landscapes and weathering studies where digital photographs have demonstrated landscape change in urban areas through time. Archival studies are integrated as part of cross-temporal research applications of photographs. Much of this part of the book is based on work performed in Oxford, UK. The second part of the book focuses on the archaeological use of photographs, comprising churchyard studies throughout England and Scotland in the UK. The scientific application of (digital) photography is presented for research scientists (landscape experts) and professional practitioners. The book is also intended for archivists interested in records of urban environments and for those readers who are interested in the geographical scope of the British cities covered in this work, including Oxford, York, Scarborough, Dunbar, Edinburgh, and Inverness. This eBook is cross-disciplinary, comprising quantitative studies within geomorphology and heritage science as well as more qualitative work in the area of historical archaeology. The photographic record contained in this book makes it a storehouse of visual records through time made accessible to a broader audience as an online (eBook) resource.

Keywords: Churchyards, conservation, cross-spatial change, cross-temporal change, digital photography, epitaphs, fieldwork, headstone introductions, headstone shape, inscriptions, kirkyards, landscape change, material culture, motifs, O-IDIP, preservation, quantitative photography, rephotography, seriation, urban environments.

This eBook provides a variety of photographic uses in the field of landscape studies. First, it examines landscape studies from a scientific perspective, including landscape change and analysis (geomorphic interpretation). The city of Oxford, UK is used as a case study for many applications in this section. In the second part, cultural (urban) landscapes are examined with examples from across Britain, from southern to northern England and Scotland. The selection of sites does not portray the country in its entirety, but provides a glimpse closer to the eastern shore, starting with Oxford and moving upwards (and east) to York, Scarborough (on the English East Coast), Edinburgh, Inverness, and Dunbar (on the Scottish East Coast). Much of the work has been performed for Oxford, England, and it will remain a focal point in this volume.

The images included here are derived from previous studies with an intensive field component. Images from Oxford were taken by the authors since 2007 and

Mary J. Thornbush and Sylvia E. Thornbush

the more recent images convey contemporary landscapes up to 2014. The images are mainly digital photographs (as evident in Chapter 3) taken with Nikon Coolpix digital cameras. However, archival images were taken before the advent of the digital camera and are of a poorer quality. Nevertheless, photo archives have been used for some part of this research and suitable digital images of these photographs are included in this eBook, starting with the next chapter.

Most of the work by the first-listed author comprises quantitative digital photography (Chapter 4). It has been published already in other books, including conference proceedings as well as academic journals. Attempts are usually made to accompany qualitative assessments alongside quantitative results for these studies. In the final section of Part I, wider applications are made of photographic use, which includes applications in the sub discipline of photo geomorphology.

The second-listed author has contributed to developing a digital record of historical archaeological monuments (headstones) at churchyards distributed towards the East Coast in England and Scotland. She approaches the topic by also focusing on urban landscapes (urbanscapes) and addresses issues pertaining to the conservation of cultural landscapes in churchyards. Part of her visual record (Chapter 6) will be contained in images presented in Part II of the eBook. Preservation issues will be considered for churchyards in urban settings in Britain (Chapter 7). Finally, she will discuss heritage sustainability for churchyards (Chapter 8), including the future (Chapter 9) of this cultural resource captured through photographs. The concluding chapter (Chapter 10) offers a synthesis of Parts I and II in the context of the contribution of photographs in understanding and capturing landscape change through time.

Part I: Photographs in Landscape Studies

Mary J. Thornbush

School of Geography, Earth and Environmental Sciences, University of Birmingham, Edgbaston, Birmingham, West Midlands, B15 2TT, England, UK

Publications exist since at least 1961 on the sub discipline of geomorphology that is 'photo geomorphology'. Its infancy was as photo geology applied to petroleum prospecting; however, it was used for a photo geomorphic study, including to discern geomorphological features, such as of tectonics and topography (Kelly, 1961). Subsequent publications included surveys of photo geomorphology (Miller, 1968), which encompassed aerial photographs in order to examine surface features (landforms). Work in India since the 1930s involved remote sensing for petroleum exploration, as for instance of the Assam Oil Company, which employed photo geological methods. Photo geomorphological studies were extended for sedimentary basins in India after setting up the Oil and Natural Gas Commission in 1956 (Talukdar, 1980). Elsewhere, in West Bengal photo geomorphologists worked with applied geomorphologists to provide soil mapping and established the 'photopedological' method at rivers (Niyogi, 1988). More work was published for Bengal Delta that involved detailed photo geomorphological studies in the West Bengal Basin (Agarwal & Mitra, 1991).

Photogemorphological mapping of coastal processes was performed for chalk cliffs between Folkstone Warren and Dover in the UK (Birch, 1990). In addition to such 'photogeomorphic' studies conducted of coastal landforms (Lind, 1974), other landscapes were also examined, including deserts (Geological Survey of India, 1982). A series of general papers appeared around this time published by the American Associate of Petroleum Geologists, which accordingly considered the direct application of photo geology and photo geomorphology to petroleum exploration (Foster & Beaumont, 1992). Several of these papers also discussed the use of photo geomorphology in geological interpretation. A lake study was conducted more recently for Sultansazligi lake in central Anatolia, Turkey that incorporated detailed photo geomorphological mapping (Erol, 1999), which was used for geomorphic interpretation of landscape development.

The number of publications in this area are somewhat limited, probably due to the speciality of the subject matter. This has included some conference proceedings, main journal articles, as well a couple of books. To the author's knowledge, the two published books on this topic are by Foster and Beaumont (1992) *Photo geology and photo geomorpholgoy*, which included several papers in his Chapter 18 on photo geology and photo geomorphology (pp. 333-551). These mainly included landform analysis, geomorphological mapping, global geomorphology, terrain analysis, aerial photographs in geomorphic studies (including structural geomorphology), geomorphology applied to oil exploration, photo geomorphic interpretations, drainage pattern analysis, and more. The other book was published more recently by Rivard (2011) and is entitled *Satellite geology and photo geomorphology: An instructional manual for data interpretation*. It contains two parts, including an examination of the data, photo geological classification of geounits, and examples of data interpretation in Part I. The subsequent part (Part II) includes details of the examples with interpretations. This includes 11 sections (with interpretations) on magmatic rocks and structures, sedimentary rocks, metamorphic rocks, geostructures, aeolian deposits and erosion forms, basinal sediments, fluvial system sediments, marine littoral systems, glacial and paraglacial geosystems, periglacial-related forms, and mass movement materials. More recently, Rivard (2011) published *Satellite geology and photo geomorphology*. Aerial photography has been expanded to include flying objects other than helicopters, *etc.* and now encompasses infrared kite aerial photography (KAP), which was field-tested at the Cheyenne bottoms Preserve of The Nature Conservancy in central Kansas, USA by Aber *et al.* (2009). These researchers were able to obtain large-scale imagery using this novel technique in aerial photography.

Miller (1968, p. 41) defined this branch of geomorphology as follows: 'photo geomorphology is the scientific study of land forms as seen on aerial photographs.' Early studies by geomorphologists like him emphasised aerial photography in particular as a way to see a variety of landscapes and measure some of their features (preferably at a small-scale of less than 1:60,000 and even more desirable at 1:20,000 (p. 43). Miller (1968, p. 52) seemed to concur that:

> Until very recently geomorphology was considered to be an interesting, but not a very important, branch of geology; by many it was even classified as more closely akin to geography than to geology. This is brought out by Tator *et al.* (1960). The vast majority of geologists, after having learned some of the basic principles of physical geology-physiography-in their elementary courses, turned their attention from land forms to such things as minerals, ores, structure, paleontology, stratigraphy or engineering or mining geology. Advanced, and especially graduate, courses in geomorphology and degrees in geomorphology,

were taken almost entirely by those few who intended to go on to teach the subject. In such a setting it is not surprising that the average geologist, a man who was later to become a photo geologist, was not particularly well grounded in the principles of true geomorphology.

As with Tator *et al.* (1960) and others, Miller (1968, p. 51) felt that 'geomorphology is the undisputed backbone of all types of photo geology.' Indeed, he felt that there was already much work in the area of photo geology within the past decade of his paper (Miller, 1968). Perhaps it is for this reason he concludes his paper by calling out for more photo geomorphological studies (Miller, 1968, p. 60):

> Our job is twofold: first, to increase the army of geomorphic data-gatherers-those technician-scientists who can create the essential penumbra of part-light; and then to increase the number, quality and ability of geomorphic photo-interpreters-the highly qualified scientists whose long and arduous labour will eventually clear away the remaining vestiges of darkness.
>
> In the umbra areas our first and most urgent need is for basic information, for reconnaissance geomorphic mapping, and for qualitative Davisian descriptions. When this phase-the penumbra phase-is well advanced, then the stage will be set for the more comprehensive interpretations, for the measurements, correlations and statistically-substantiated explanatory hypotheses. Only in this way will geomorphology be able to play the important role which it surely is destined to play.

The use of 'interpretation' was used differently from its currently use for what Miller (1968, p. 49) thought specifically as follows: 'Interpretation, then, can be properly restricted to mean the process of giving meaning to things which have been identified.' Unlike the geologists of his day, he considered various landscapes to be worthy of photo geomorphological attention, including rocks and geological structures, climate, weathering (especially chemical and/or mechanical), erosion, deposition, soils, and vegetation (p. 50). His application was, therefore, not as restricted as that of photo geologists, who focused on petroleum exploration geomorphic interpretation (as in Foster & Beaumont, 1992). Photo-interpretation includes some important work in this area. For instance, Kellerhals *et al.* (1976) provided a classification of fluvial features used in conjunction with photo interpretation or field reconnaissance which included valley wall, valley flat and channel features, codifications of channel patterns addressing meander bend regularity, islands, and channel bars, and different classes of channel lateral activity. At around the same time, Mollard (1973) and later Mollard and Janes (1984) published a couple of works that addressed airphoto interpretation.

Even though aerial photography is pivotal to this subdiscipline, other approaches should be recognised within photo geomorphology. One of the chief ambitions of the first part of this eBook is to make a contribution to the development of photo geomorphology as a photo interpretation approach for the examination of changing landscapes from the ground rather than from the air (from air-borne or orbiting objects, such a satellites), which is arguably better-suited to GIS applications. It is time to also develop the use of close-range ground-based methods in photo geomorphology. These can be employed quantitatively indoors and in the field by physical and human geographers (Sidaway, 2002), which is something that has been neglected by studies that have employed photographs as slides in lectures for demonstration of landforms (for instance, McKendrick & Bowden, 1999, 2000). Photo-based quantification is rarer than the use of photographs qualitatively in the pictorial depiction of landscapes (Swallow *et al.*, 2004). Some examples already exist, however, that denote how photographic input can be used, in this case, in the laboratory to accumulate visual evidence of etch features on marble at a range of lighting angles, which was effective when the angle of incident equals the angle of reflectance (Mottershead *et al.*, 2003).

By way of introduction, therefore, it is necessary to layout the approach taken in the current volume. Most importantly is the recognition that landscape studies belong under the auspice of geomorphology as the study of surficial processes and landforms (Thornbury, 1970). Second, and equally significant, is the recognition that aerial photographs (air photos) are not the sole valuable scale for a photo geomorphologist to adopt. This is evident in published studies by the first-listed author as well as others who engage in modern photo grammetry (for instance, Chandler, 1999, who advocated a close-up approach for geomorphological studies). Even though photo grammetry has been developed (published in journals, such as the *ISPRS Journal of Photo grammetry and Remote Sensing*), by comparison, photo geomorphology remains an undeveloped branch of geomorphology that is scattered in studies appearing in various publications as a type of applied geomorphological technique within landscape studies and geomorphology in the study of surface processes and landforms. Indeed, in this volume it will become apparent that close-up photography is useful to capture small details of landscape change, such as at the micro scale for weathering studies (Viles, 2001). The approach taken here is that detailed (digital) photography can be used quantitatively as well as qualitatively to capture change in the landscape, including in the built environment in urban settings. This capturing of cross-temporal change can help to establish a digital record of the current state or condition of monuments, from historical buildings located in cities to headstones found in urban churchyards.

In the this first part of this eBook, photography, and digital photography more specifically along with repeat photography (rephotography) and quantitative re/photography, will be introduced and case studies provided to exemplify application. Three case studies contained in this first part address the archival record (photo archives) and the cross-temporal pictorial depiction of historical buildings in the Oxford city centre. Another case study in this first part conveys the close-up capturing of bioweathering and its agents in central churchyards. This is followed by case-study content of a quantitative nature, measuring lightness and colour at a string course located along a (polluted) main roadway in central Oxford.

ACKNOWLEDGEMENTS

Declared None.

Chapter 2: The Archival Record

The Archival Record

Abstract: This chapter examines the photoarchives available in print format (as at the archives of Oxford colleges) as well as online (electronic versions) of the record. Archival material is crucial in cross-temporal research in order to extend back the temporal framework as far back as possible in the historical record, in this case using historical photographs in various available formats. The archival record is considered here from a photographic perspective, with consideration of photographs and how they have been used to track cross-temporal landscape change. These historical photographs have sometimes appeared as postcards in the archival record or as digital images, and can be the basis of before-and-after comparisons of change. This chapter contains a case study that examines how modern photography can stand with historical photographs in order to portray cross-temporal change of various buildings located in central Oxford.

Keywords: Before-and-after comparisons, cross-temporal change, online databases, photographs, pictorial representation.

Photographs are an ideal medium to assess landscape change, as they are realistic visual depictions of landscapes. Until recently they have been used more descriptively and interpretatively than analytically. For instance, Ryan (1997) captured what was conceived as the 'visual truth' of British imperialism in Africa. Unlike other artistic forms (drawings and paintings), which can distort the representation of landscape, photographs allow for objectively capturing landscape change. It has been recognised, however, that photographs can be altered and argued that other artistic records may convey more accurate information. Sontag (1977), for instance, posited that photographs are just as constructed as paintings because of issues associated with selection. Others, such as Ode *et al.* (2010), were concerned with representative sampling in photography. Bass (2004) presented several caveats of using photographs to track landscape change, among these are problems associated with their representativeness and compositional bias. This can be overcome, however, through the use of rephotography, although this has its own issues, such as that of hidden time between photographs, which must be inferred and thereby necessitating interpretation. Nevertheless, this methodology (of rephotography) has been employed by many researchers in the natural sciences in order to capture landscape change.

Photographs can be employed to decipher long-term environmental change. Photoarchives, for example, are a repository of historical photographs that have much potential for incorporation into repeat photographic (rephotographic) surveys. Thornbush and Viles (2005), for instance, were able to extend the temporal record back for century-scale capture of landscape change for an urban environment at Magdalen College, Oxford. These archival photographs were used in conjunction with other records obtained in the archives of Magdalen College, with traffic records conveying a recent decline in traffic congestion at Magdalen Bridge since the Oxford Transport Strategy (OTS) implemented in June 1999. Based on photographic evidence, the façade of Magdalen College appeared blackened before motor traffic posed any real threat. However, traffic may be responsible for façade darkening at low levels that is noticeable in recent photographs. This study showed that cross-temporal analyses, although useful, can be hampered by scale differences limited by the purpose of initial photography. Seeing soiling from a distance in the photoarchival record makes it difficult to quantify change and only qualitative assessments may be made; this is a similar impediment when addressing decay features, which need to be visualised at a small scale. Sampling bias also limits photographic collections. Moreover, colour-based assessments, as for instance of biological weathering, are not possible based on black-and-white photographs. There are also lighting inconsistencies in photoarchival records because of the different times when photographs were taken. Clouds could also complicate lighting conditions. Furthermore, the camera used and the angle at which photographs are taken affect comparisons. Nevertheless, despite these limitations, photoarchives enable the capturing of longer time scales in monitoring. Thornbush and Viles (2005) conveyed the usefulness of archival materials to extend records of building soiling and decay to the century timescale. This was impeded, however, by the lack of close-up photography, but did manage to portray some basis for evaluation since the late 19th century and up to the late 20th century. The authors encouraged further research to address lighting conditions in outdoor photography and to develop photographic calibration for quantitative assessments.

American researchers, in particular, have tracked landscape change in mountainous environments most notably. For instance, Butler (1994; Butler & Malanson, 1990, 1993) is recognised for his use of rephotography to assess change in the northern Rocky Mountains in Montana. Most recently, he published with DeChano (Butler & DeChano, 2001) using postcards to track landscape change. Others have employed a similar approach at other (mountainous) locations, including the Rocky Mountains in Colorado and in the San Juan Mountains (Elliott & Baker, 2004; Manier & Laven, 2002). Landscapes photographs taken since 1917 near Whiskey Mountain in the Reynolds Creek Experimental Watershed situated in southwestern Idaho were subject to a quantitative analysis by Clark and Hardegree (2005) in their assessment of long-term dynamics in vegetation cover in rangelands. Rephotography as a methodology that made it possible to analyse landscape change in Vermont between 1860 and 1990 in order to discover enhanced erosion with the removal of a vegetative cover (Bierman *et al.*, 2005). Another similar study of long-term vegetation change took place in the Oklahoma Cross Timbers also in the USA by Griffin *et al.* (2005).

A classic study (Webb, 1996) conducted at the Grand Canyon, USA compared landscape photographs taken as part of an expedition in 1889-1890 and compared these with more contemporary images. This has been followed up more recently by the author in an edited volume on *Repeat photography: Methods and applications in the natural sciences* (Webb *et al.*, 2010). This book commences with a history of repeat photography in the introduction and then continues by relaying techniques; with applications in the geosciences, including for the measurement of glacial retreat, and population ecology as well as ecosystem change, plus cultural applications before a consideration of the future of the methodology by the editors. The authors refer to various online archival records available for the USA in their Fig. **1** (p. 13). Others, such as Hall (2001, 2007), have employed 'photomonitoring' in ground-based photographic monitoring. In addition to the use of such oblique ground-based images, as already conveyed in the Introduction, vertical photography based on aerial photographs has been used to classify landscapes at various resolutions. There are various examples of this approach, including Antrop and van Eetvelde (2000); Clark and Hardegree (2005); Dahlberg (2000); Hudak and Wessman (1998); Kadmon and Harari-Kremer (1999); and Rango *et al.* (2005).

Bennett *et al.* (2000) used close-range vertical photography to measure cover changes in grasslands and found it to be both accurate and precise. In comparison with aerial photographs and satellite images, historical and recent landscape photographs were found to provide a higher resolution and deeper historical reach and useful to illustrate change (Kull, 2005). Other grasslands were examined by Burnside *et al.* (2003), who examined landscape change in the South Downs of Sussex, UK based on photographs taken between 1971 and 1991. Swetnam *et al.* (1999) similarly established a sequence of progressive tree invasion in a montane grassland restoration project located in northern New Mexico in the 20th century using aerial photographs. Brook and Bowman (2006) used aerial photographs to convey the expansion of forest that occurred in the past 5 decades in the Australian monsoon tropics. The methodology has also been applied to monitor change in cultural landscapes, as for instance by Nüsser (2001).

In recent years, there has been a proliferation of landscape rephotographic studies that have been published from climatic and biological studies. Kullman (2006), for instance, used this methodology in order to convey a causal association between temperature rise and biotic evolution. Historical photographs were also the basis of an assessment by Munroe (2003) of the biotic response to climatic warming in the northern Uinta Mountains, USA in the past 130 years or so. A study took place by Hereford *et al.* (2006) during the wet period from about 1976-1998 in order to assess the size and density of bush species. The methodology was also employed alongside (site-specific) meteorological measurements in order to monitor phenologic events in plants by Crimmins and Crimmins (2008). Thousands of images were used by Proulx and Parrott (2009) in order to examine structural complexity at 72 sites in an old-growth forest. In some cases, digital photographs are employed in rephotographic studies. For instance, digital photography was combined with digital image analysis to study intertidal benthic organisms, as by Pech *et al.* (2004).

In addition to the use of photography in the physical sciences, human geographers have also attested to its importance for description as well as analysis of the historical record (Rose, 2000, 2008). However, the application within the physical sciences has flourished, particularly in the geosciences, as in geomorphology (see Lane *et al.*, 1993). For instance, an archive-based study based at Trinity College, Oxford used photographs to quantify changes in the size of blistering evident on old limestone walls apparent in backdrops of group photographs from the 19th century (Viles, 1993). Others have explored photoarchival records for other reasons (other than studies on stone soiling and decay), such as Foote (1985), who examined changing architectural features found in the urban landscape of Austin in Texas, USA.

Archival photographs have only recently been employed in research into environmental impacts on historical buildings. Work by Viles (1994) and other researchers, such as Davidson *et al.* (2000) referred to the archival record in the monitoring of soiling patterns on limestone building façades since the early 19th and 20th centuries in Oxford, UK and Pittsburgh, Pennsylvania, USA, respectively. Camuffo and Bernardi (1993) considered marble soiling from a baseline study by Becatti from 1957 that compared the condition of the Aurelian column with 35-year-old restoration photographs. Andrew (2002) incorporated photographic pairs (photo-pairs) into his research into the perception of soiling and age estimation based on comparisons of architecturally similar buildings before and after cleaning. Studies into stone decay have tended to be rather shortsighted, with a tendency of short-term observations up to 3 years (Smith, 1996). There is a need to extend the temporal framework of the current understanding of the weathering of historical building stone and the archival record, as through photoarchives, has much potential for the derivation of decadal- to century-scale evaluations of stone soiling and decay.

The book *Oxford stone restored* (Oakeshott, 1975) incorporated the use of repeat photography (rephotography) in the cross-temporal representation of some 17 Oxford buildings, mostly Oxford colleges, conveying landscape change at the façade scale. It denotes the work between 1957 and 1974 of the Oxford Historic Buildings Fund, containing archival photographs as well as restoration images taken by J. W. Thomas. These images provide before and after glimpses of the appearance of building exteriors since restorations commencing in the 1950s. Thornbush (2010b) took a follow-up rephotographic survey of a selection of building façades and, using this evidence, derived an inventory of weathering forms. The results revealed that over time buildings become darker at the top, which could be attributable to top-down climatic exposure as well as air pollution and the settling of particulates. The weathering features that were identified in this study included encrustation, blistering, scaling, and flaking as well as granular disintegration and pitting.

Further archives were visited at Oxford colleges, including Trinity and Pembroke colleges in a study of urban greening. This work has made a unique contribution to the field of historical rephotography in a built landscape derived from photoarchives. Here, the employment of archival materials enabled the establishment of an environmental history of the use of ivy and creeper in central Oxford since 1861 (Thornbush, 2013c). Their growth appears to have been well-maintained on the Chapel Quadrangle at Pembroke College, where climbing plants were kept away from windows and off roofs in the late 19th century. Nevertheless, it seems that around 1953 climbing plants were cleared from walls, which in the 19th century had been backdrops included in group photographs. There is, however, some indication that ivy- or creeper-clad building backdrops were favoured, perhaps suggesting a continued social preference for this visual aesthetic.

Postcards are derived from historical photographs and these can represent the landscape through time. Butler and DeChano (2001), for instance, employed postcards in their study of landscape change. Similarly, discarded postcards were examined in an article by Thornbush (2008a), which remained intact as representations of historical photographs. She examined postcards contained in a book *Pictures in colour of Oxford* (c. 1907) in order to establish a record of environmental change of the greenness of Oxford colleges, including All Souls, Balliol, Brasenose, Corpus Christi, Exeter, Jesus, Lincoln, Magdalen, Mansfield, Merton, New, Oriel, Pembroke, Queen's, St. John's, Trinity, Wadham, and Worcester, since the early 20th century. The postcards conveyed the growth of climbing plants (ivy and creeper) growing on the exterior of Oxford buildings between 1903 and 1923, with the possibility of earlier photography. Some colleges, including Worcester and New colleges, maintained their use of climbers; this was not the case for all colleges, such as Magdalen and St John's. Colleges with the least cover of climbers included All Souls, Christ Church, and Corpus Christi. These climbing plants were kept in back areas of some of these colleges, such as at Christ Church, in gardens and/ or on boundary walls, as at Trinity College. The front façades of buildings were often clear of climbing plants. Some colleges allowed these climbers to grow freely, where they were visible growing onto roofs and towers, as for instance at All Souls, Magdalen, New, and Trinity colleges. Other colleges, such as Christ Church and Worcester College cut back climbers in order to prevent them growing too high on buildings. From the images depicted in postcards, it is sometimes possible to ascertain the season when the photograph was taken and, therefore, establish the type of climbing plant (because ivy is an evergreen and creeper is deciduous). Creeper was visible, for instance, at Oriel College and Corpus Christi (as well as Balliol, Trinity, Exeter, and Lincoln colleges).

It is possible to employ online databases, in addition to printed media, to investigate cross-temporal change of landscapes. Thornbush (2013a) most recently used historic photographs derived from ViewFinder from English Heritage (2011) to study the appearance of urban greening in central Oxford. The vegetation comprised in this study included climbing plants (ivy and creeper) evident in photographic collections contained with this online database. The Henry W. Taunt (1860-1922) collection was particularly useful in this case and a total of 1,123 images were categorised as either containing ivy and creeper or being free of these climbers. This vegetation cover was investigated both at Oxford colleges and non-University buildings in the city centre. The findings reinforce that climbing vegetation was popular during Victorian times towards the end of the 19th century, particularly in the 1880s. Even though it was not possible to obtain close-up photographs, images at the building scale were of a sufficient detail to provide information of the presence/ absence of a vegetation cover on buildings. Several problems were encountered, however, concerning image and database quality.

The following case study on the historical exposition of historical buildings in central Oxford draws on research that employs historical photographs to track the use of climbing plants on building façades. The online database ViewFinder from English Heritage is used in a study of important non-college buildings (some of which still pertain to the University of Oxford). Using this approach, it is possible to establish a cross-temporal comparison of a vegetative cover of climbing plants on such important buildings in the city centre, including the Clarendon Building, Sheldonian Theatre, Radcliffe Camera, and the Ashmolean Museum.

Case Study: A Historical Exposition of Heritage Buildings in Central Oxford, UK

Abstract: This case study examines environmental change associated with climbing plants (ivy and creeper) on several heritage buildings in central Oxford using historical photographs. ViewFinder from English Heritage was used to access the photoarchives pertaining to non-college buildings in the city centre. The selected buildings investigated in this study included the Clarendon Building and Sheldonian Theatre, Radcliffe Camera, the University Church of St Mary the Virgin, St Michael's church, and the Ashmolean Museum. The findings reveal that most of these buildings have been a showcase as heritage works and were not, historically, covered by ivy and/ or creeper growth. Possible reasons are considered for this, including the importance of keeping them well-maintained and clear of climbing plants that would otherwise hinder the appreciation of their architectural splendour. Importantly, it is improbable that these findings are a product of an incomplete photographic record.

INTRODUCTION

An effort is underway by the author to investigate landscape change through an environmental history of climbing plants or climbers (ivy and creeper) apparent historically at Oxford colleges based at the University of Oxford, UK. These studies examine college archives in conjunction with other historical records, including photographic collections. The purpose of this research is to establish a clearer understanding of the use of ivy and/ or creeper in an attempt to ascertain the reason for their appearance through temporal sequences of change captured by photographs taken at various colleges. For an exhaustive study, it is also necessary to examine the non-college record of historical buildings located in central Oxford. It is important to do this in order to establish whether certain trends are apparent at the colleges that do not include non-college buildings. Perhaps the use of climbing plants was particular to Oxford colleges and not extending to non-college buildings at the University of Oxford.

There are no precedent (known) published studies examining ivy and/ or creeper through the archival record, except for recent works by the author. The most relevant published work is a recent research paper by Thornbush (2013b), where she conveyed findings of a search in ViewFinder for all buildings located in central Oxford that were covered in ivy and/ or creeper. She also considered these buildings when they were not covered in climbing plants and used this record to establish an environmental history of the appearance of climbers on a variety of buildings situated in central Oxford. Her findings show that most buildings (53%) were covered in ivy and/ or creeper and also were covered in the 1880s and 1900s.

The author encouraged further study of Oxford colleges; however, this is not the emphasis of this case study. This was already considered in a previous study by Thornbush (2013c), where the author examined the cross-temporal appearance of climbing plants at two Oxford colleges, with case studies at Pembroke and Trinity colleges. She discovered climbers on buildings since the 19th century (with records from 1889) up until 1953, when they were cleared at Pembroke College; the record denoted a similar temporal range for Trinity College between 1861 and 1964. Thornbush (2008a) also examined ivy and/ or creeper on buildings in central Oxford, this time based on a printed book entitled *Pictures in colour of Oxford* (c. 1907). This research revealed that it was possible to establish an environmental history based on postcards, with Oxford colleges appearing green in the early 20th century. Of the colleges examined, including All Souls, Balliol, Brasenose, Corpus Christi, Exeter, Jesus, Lincoln, Magdalen, Mansfield, Merton, New, Oriel, Pembroke, Queen's, St. John's, Trinity, Wadham, and Worcester, those colleges with the least cover of climbing vegetation were All Souls, Christ Church, and Corpus Christi.

Prior studies included an archival study conducted at Magdalen College, which examined the photo archival record in combination with historical records of traffic congestion at Magdalen Bridge (Thornbush & Viles, 2005, pp. 40-57). The authors, however, were more interested in investigating stone soiling and decay along the High-Street frontage in connection with traffic pollution rather than the use of climbing plants, as in the current study. Nevertheless, they did include a photograph from the late 19th century[1] that clearly portrayed ivy and/ or creeper on Magdalen Tower. Though archival studies have been conducted using historical photographs from Trinity College (Viles, 1993, pp. 308-326)[2] and even of the Ashmolean Museum (Viles, 1994, pp. 1-27)[3], they focused on stone decay in general rather than on biological impacts.

The interest in the evidence of climbing plants stems from current feelings of gardeners, which tend to be disparate, either for or against the use of ivy in particular. At Trinity and Pembroke colleges, for instance, the current practice is to avoid use of ivy. In fact, looking around much of Oxford, it is difficult to find evidence of climbing plants, except perhaps on boundary walls (*e.g.*, at Trinity and Worcester colleges) or unmaintained remains (*e.g.*, at Rewley Abbey and the Old City Wall). Nevertheless, some colleges, such as Christ Church, Exeter, and Lincoln, have established creeper on (soiled) building façades. There is a current interest in the use of ivy, however, and English Heritage funded a project (between 2006 and 2009) that investigated ivy on walls.[4] This project considered the maintenance of ivy and the current practice of gardeners at colleges and at other locations in England. The study did not, however, consist of an in-depth investigation into the historical record for an environmental history of the appearance of ivy that may make an important contribution to an informed understanding of environmental change.

The purpose of the current study is to track the use of ivy and/ or creeper on walls in Oxford using historical photographs from the archive ViewFinder maintained by English Heritage. Non-college heritage buildings pertaining to the University of Oxford were selected for a consideration of use and practice in central Oxford outside colleges. Such an approach makes for a more complete understanding of environmental history outside the college domain and may help to isolate trends in the record that may not be apparent by studying only college properties.

This investigation was conducted as follows. Historical photographs, mostly from the Henry W. Taunt collection (1860-1922), were searched for in an image gallery in ViewFinder in the English Heritage domain.[5] A search of the image gallery was conducted with the name of the building as the place name, in the county of 'Oxfordshire', in the area of 'Oxford'.[6] The buildings searched in this database were: 1) the Clarendon Building and Sheldonian Theatre, 2) Radcliffe Camera, 3) the University Church of St Mary the Virgin, 4) St Michael's church, and 5) the Ashmolean Museum. A map of where these buildings are situated in central Oxford appears in Fig. **1**. Only photographs depicting the outside (or exterior) of these buildings were examined in this study. Specific attention was paid to any climbing plants visible on the exterior of the buildings, but photographs missing these were also considered in this investigation.

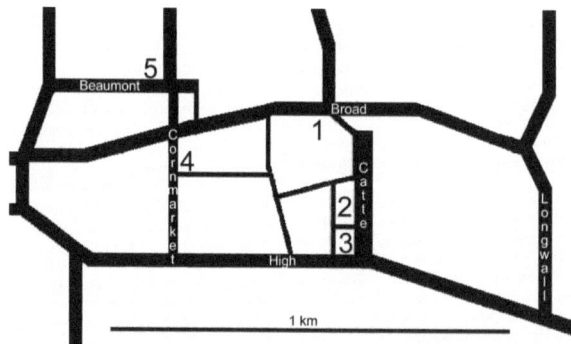

Figure 1: Map of the heritage buildings located in central Oxford examined in this study, including 1) the Clarendon Building and Sheldonian Theatre, 2) Radcliffe Camera, 3) the University Church of St Mary the Virgin, 4) St Michael's church, and 5) the Ashmolean Museum.

Historical Photographs of Heritage Buildings

A photograph from the Taunt collection taken in 1870 (HT01919)[7] shows an encrusted façade, seen facing north from Broad Street, with evidence of exfoliation along the bedding planes of the slabs of limestone, as for example on its columns (Fig. **2a**). There is no appearance of ivy and/ or creeper on the building façade in this photograph. A subsequent photograph from 1875 (CC72/02252) depicts the blackness of the building façade, which is again devoid of any plant cover. A third image, an illustration[8] taken from the same vantage point as the previous two, from 1907 (CC50/0070) also fails to capture any evidence of ivy and/ or creeper on the building façade.

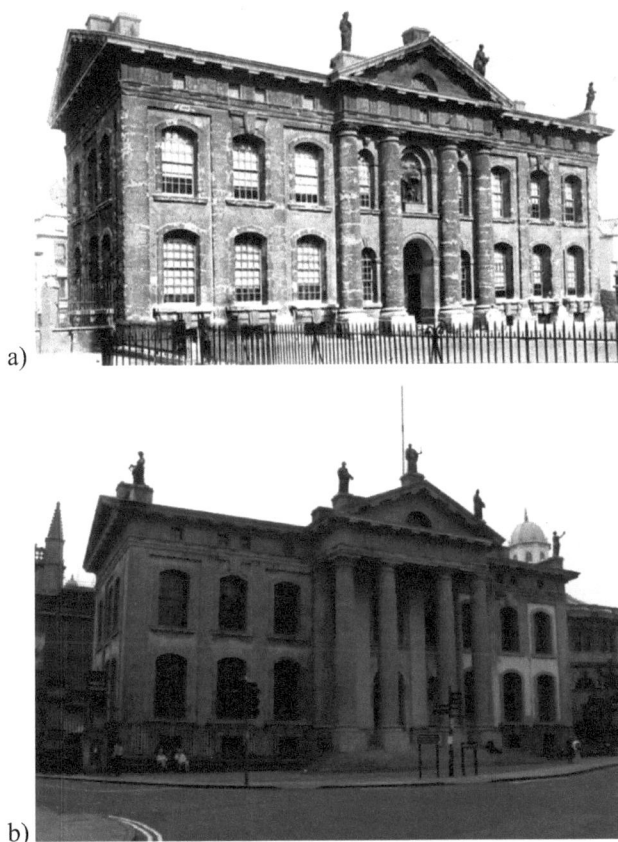

Figure 2: Image of the Clarendon Building (designed by Nicolas Hawksmoor for the Oxford University Press, completed between 1711 and 1715) from Broad Street taken in a) 1870 (HT01919) (from the ViewFinder collection of Henry W. Taunt, 1860-1922; reprinted with permission from Oxford County Council – Oxfordshire History Centre) and b) in the spring of 2007 (taken by M. Thornbush).

In the Taunt collection between 1860 and 1922 (CC50/00406), both the east-facing side of the Clarendon Building and the Sheldonian Theatre seen from Catte Street appear clear of climbing plants. This is also true for 1875 (CC50/00407), but not for 1885 (CC50/00405), where a thick growth of ivy and/ or creeper is evident growing on the back of the Clarendon Building that can be seen from Catte Street (Fig. **3a**). The Sheldonian Theatre, on the other hand, remains clear of any growths. Another photograph taken in the same year, in 1885 (CC50/00403), looking eastward from Broad Street fails to show a similar growth of climbing plants towards the other side of the Clarendon Building. This was also the case from this perspective five years earlier in 1880 (HT03435), as well as in later years.[9]

a)

b)

Figure 3: Ivy and/ or creeper growing on the back of the Clarendon Building, also showing the Sheldonian Theatre (commissioned by Chancellor Gilbert Sheldon and completed in 1669), as seen from from Catte Street in a) 1885 (CC50/00405) (from the ViewFinder collection of Henry W. Taunt, 1860-1922; reproduced by permission of English Heritage) and b) in the spring of 2007 (taken by the M. Thornbush).

Images from the ViewFinder gallery of the Radcliffe Camera consistently portray an exterior clear of any climbing plants. Photographs from the Taunt collection[10] all indicate a cleared exterior (Fig. **4a**). A print and some engravings, in particular, fail to convey the texture of the middle section of the building and also fail to convey the soiled state of its bottom portion. This encrustation on the building is evident in more recent photography from the Eric de Mare collection from 1945-1980 (*e.g.*, AA98/04809).

a)   b)

Figure 4: The Radcliffe Camera (completed in 1748, from funds remaining from the will of John Radcliffe, and designed by James Gibbs) in a) 1865 (HT00732) (from the ViewFinder collection of Henry W. Taunt, 1860-1922; reprinted with permission from Oxford County Council – Oxfordshire History Centre) and b) in the spring of 2007 (taken by M. Thornbush).

On its south side, on its frontage onto High Street, University Church of St Mary the Virgin is a porch that has a history of changing ivy and/ or creeper cover. During the time of the Taunt collection of 1860-1922 (CC49/00269), the frontage of the church appeared well-vegetated, with some climbing plants evident amongst the greenery.[11] A close-up image of this photograph (CC49/00273) better demonstrates the amount of ivy and/ or creeper on the porch. A photograph dating to 1875 (CC49/00270), for instance, portrays ivy and/ or creeper growth on the south face of the building east of the porch. The greenery of the south-facing façade of the church is noticeable as one looks westward on High Street, as in 1885 (CC72/02059). It is possible that it was creeper on the south façade of the church, with its visibility reduced in winter months.[12] Much of what may be creeper can be seen growing thickly in 1890 (HT05137) (Fig. **5a**); unless it was cleared between 1890 and 1898, then it is likely creeper (a deciduous plant), rather than ivy (an evergreen plant), adorning the south-facing front of the University Church of St Mary the Virgin. A photograph from 1896 (HT07728) helps to substantiate that it is creeper growing on the High-Street frontage of the church (Fig. **6**). Here, the magnolia tree, which is also a deciduous plant, can be seen without leaves; whereas, leaves are visible on it in the photograph from 1890.[13] At the back of the building, a photograph from 1909 (CC49/00280) depicts ivy and/ or creeper growing on the wall at the back of the building, facing onto Radcliffe Square. On 15 June 1910 (OP03797)[14] strands of creeper are still visible on the High-Street frontage of the church.

Figure 5: The University Church of St Mary the Virgin with its Baroque-style south porch (built in 1637 probably by Nicholas Stone) shown from High Street in a) 1890 (HT05137) (from the ViewFinder collection of Henry W. Taunt, 1860-1922; reprinted with permission from Oxford County Council – Oxfordshire History Centre) and b) in the spring of 2007 (taken by M. Thornbush).

Figure 6: The University Church of St Mary the Virgin on its south side along High Street probably taken in winter, since only the stems of creeper are visible on its façade in 1896 (HT007728). (From the ViewFinder collection of Henry W. Taunt, 1860-1922; reprinted with permission from Oxford County Council – Oxfordshire History Centre).

St Michael's church can be seen from a distance on Cornmarket Street because of its tower[15]. The Tower is portrayed in the Taunt collection between 1860 and 1922 (HT00118), when some stems may be visible on its south face. This is not clear here, and a photograph from 1885 (CC72/00791) shows only a small climbing plant on the lower portion of the Tower (Fig. **7a**). Its north face is clear, as evident in 1907 (HT10442), although what appears to be moss is growing on the west-facing side of the Tower. This growth is more easily visible in 1909 (HT11127), perhaps because it has grown in the 2 years.

Finally, at the Ashmolean Museum, in 1880 (CC50/00336) there is no appearance of ivy and/ or creeper on the south-facing façade of the building. In 1885 (CC50/00335), looking west along Beaumont Street, the building is still devoid of any climbing plants (Fig. **8a**). The Ashmolean appears much as it does today in the Eric de Mare collection from 1945-1980 (AA98/04780). This building does not seem to have much of a history involving ivy and/ or creeper growth.

All these heritage buildings in the present-day do not have any ivy and/ or creeper cover on them. This is evident in several modern photographs taken by the first-listed author appearing in Figs. **2b-5b** and **7b-8b**. It is clear that the current practice is one of clearance, even on some of the heritage buildings that previously had climbing plants (*e.g.*, the façade of the University Church of St Mary the Virgin in Fig. **5b**).

a)

b)

Figure 7: St Michael's church from Cornmarket Street in a) 1885 (CC72/00791) and b) in the spring of 2007 (taken by M. Thornbush), with its late Anglo-Saxon tower dating to the late 11th century (the rest of the church is from the 13th century or later). Ivy and/ or creeper can be seen growing up a small part of the south-facing side of the tower. (From the ViewFinder collection of Henry W. Taunt, 1860-1922; reproduced by permission of English Heritage.)

a)

b)

Figure 8: The Ashmolean as seen looking west onto Beaumont Street in a) 1885 (CC50/00335) and b) in the spring of 2007 (taken by M. Thornbush), devoid of climbing plants. (From the ViewFinder collection of Henry W. Taunt, 1860-1922; reproduced by permission of English Heritage).

CONCLUSION

Oxford is a very photogenic city and its photographic record provides a detailed source of information about its heritage buildings, making it possible to assess environmental change. Ivy and/ or creeper can be seen growing on non-college buildings, including on the back of the Clarendon Building, the façade of the University Church of St Mary the Virgin, and the south-facing side of St Michael's Tower. Other non-college buildings, however, have no photographic record of climbing plants; this includes the Sheldonian Theatre, the Radcliffe Camera, and the Ashmolean Museum. These are substantial buildings, which possibly did not have any ivy and/ or creeper growth because they were well-maintained historically, as into the present-day. It is true that ivy, for instance, is often associated with neglect. So, for this reason, perhaps buildings, such as the Sheldonian, Radcliffe Camera, and the Ashmolean, were never green in order to avoid any incidental damage that could have been brought on their façades from climbing plants, in an effort to not risk marring these architectural monuments.

The Sheldonian Theatre and Radcliffe Camera were particularly blackened at one point.[16] So, it is likely that the use of ivy and/ or creeper was not to cover-up the soiling, or even decay, of buildings exposed to air pollution.[17] Had this been the case, climbing plants would have been used on such badly soiled exteriors. Instead, it is argued here that neither ivy nor creeper were allowed to grow on these buildings because of their architectural splendour so that passersby could appreciate their lines of contour and definition. Climbing plants would have severely hindered this and not particularly complemented the architectural design.

At the Clarendon Building, climbing plants were only allowed to grow towards the back region of the building well away from its façade onto Broad Street. There is photographic evidence of a similar approach being taken, for example, at Christ Church, University, All Souls, and Balliol colleges. Here, the frontages of colleges were kept clear of any climbing plant growth, although the sides of the main buildings and other parts in quadrangles may have housed some ivy and/ or creeper. At Exeter College, for instance, creeper was allowed to grow in the main quadrangle, but the frontage of the College facing onto Turl Street was kept clear. Another point regards the use of creeper.[18] It is clear that creeper was used rather than ivy, which is an evergreen. A possible reason for this, especially at the porch of the University Church of St Mary the Virgin,[19] is to avoid completely hiding architectural features of the building year-round. Creeper is also a way of adding autumn colour to the exterior of buildings, contributing to a pleasant aesthetic.

Finally, the completeness of the historical record must be considered. It is often the case that photographs are sold or lost because of copyright by newspapers, for postcards, *etc.* In other cases, the families of former students donate albums to colleges for their archival collections. The Trinity College archives, for instance, include photographs taken at other colleges, such as Magdalen and Worcester colleges. Similarly, the archives at Pembroke College also include photographs of other colleges, such as of Queen's, Magdalen, Worcester, St John's, Christ Church, Oriel, New, All Souls, Trinity, and Balliol as well as non-college buildings, such as the University Church of St Mary the Virgin, the Sheldonian Theatre, the Clarendon Buildings, and the Radcliffe Camera. In the Pembroke College archives, ivy and/ or creeper was spotted on Magdalen Tower around 1890-1895 and around 1889-1894 on the façade of the University Church of St Mary the Virgin, at the back of the Clarendon Building towards Catte Street, at the Cloisters of Magdalen College and on its tower, at New College from the gardens, and on the gate in the front quadrangle of All Souls College. Photographs depicting the Sheldonian Theatre and Radcliffe Camera (also in the Pembroke College archives) do not show any evidence of ivy and/ or creeper growth between 1889 and 1894. This suggests that the photographic record is not failing to show a complete record of environmental change, but that rather these buildings were kept clear of climbing plants.[20]

NOTES

1. Adopted from c. 1896-1899, in their Fig. **3**.

2. She examined blistering based on archival photographs, including group photographs such as of rowing eight from the summer of 1882.

3. She included some case studies of the Bodleian Library from the Catte Street front in 1904, 1939, 1952, and 1993 as well as the Tower of the Five Orders in c. 1880, c. 1900, 1956, and 1993; the Ashmolean Museum in 1882, 1960, 1963, 1976, and 1993; and Wadham College from the Parks Road front in c. 1820, 1885, 1956, early 1960s, and 1993. Here, Viles (1994) saw the potential of archival evidence, especially photographs, providing an objective framework for the study of the history of stone decay.

4. The 'Ivy on walls – biodeterioration or bioprotection?' project. More details can be obtained from the webpage: http://www.ouce.ox.ac.uk/research/arid-environments/rubble/projects.php.

5. ViewFinder, http://viewfinder.english-heritage.org.uk/home.asp?JS=True, 10 April 2007.

6. A simple search resulted in seven images for 'Clarendon Building', 18 images for 'Sheldonian Theatre', 11 images for 'Radcliffe Camera', 49 images for 'St Mary the Virgin Church', 6 images for 'St Michaels Church', and 28 images for 'Ashmolean Museum'.

7. Reference numbers appear for each image in brackets after the given date, in this style, throughout this case study.

8. It is interesting to note how clean the building appears in this illustration that evidently does not accurately portray the blackening of the façade apparent in the previous photographs.

9. Based on engravings from 1895 (CC72/02255) and 1907 (CC72/02253).

10. From 1860-1922 (HT01702), 1865 (HT00732), 1901 (HT08485 and HT08486), a print dating to 1907 (HT10096) as well as images of engravings also taken in 1907 (HT10137, HT10622, and HT10648).

11. This must have varied quite a bit during this time, because in 1870 (CC49/00268) most of it was gone. The ivy and/ or creeper must have been cleared in 1870, for an image shows ivy and/ or creeper on the porch in 1870 (HT00746), and another from the same year (HT00895) fails to capture climbing plants.

12. As, for instance, in 1898 (CC49/00272) when non-leafy material is evident on the front of the building in vicinity of the porch.

13. An engraving from later in 1907 (CC49/00266) fails to show creeper growth at the front of the building.

14. From the Potter collection.

15. This refers to St Michael's Tower, which is supposed to be Saxon.

16. As, for instance, depicted in an image from 1880 (HT03435) of the former building and in 1865 (HT00732) as well as 1901 (HT08486) of the latter building.

17. From coal burning in the city centre (cf. Viles, 1996, pp. 359-372) or traffic pollution (cf. Antill & Viles, 1998, pp. 28-42).

18. As, for instance, at the University Church of St Mary the Virgin (and also, possibly, the tower of St Michael's church).

19. As evident, for instance, in 1898 (CC49/00272).

20. Though it is always possible that some records can still be uncovered that will reveal a different story.

ACKNOWLEDGEMENTS

Dr. Thornbush is grateful to the staff at the Centre for Oxfordshire Studies for assisting her during her search for historical photographs and for suggesting ViewFinder. The historical photographs have been reproduced by permission of English Heritage and Oxford County Council – Oxfordshire History Centre.

Chapter 3: Digital Photography

CHAPTER 3

Digital Photography

Abstract: Digital photography is the form photographic form that is most easily employed in the field and for quantitative study of landscape change. Digital cameras are increasingly cheaper and images more capable of capture detail. For these reasons mainly, digital photography is integrated into research studies both qualitatively (pictorially) or for quantification (measurement). Entire databases are being constructed using digital cameras, including museum collections, many of which now have a digital database or gallery associated with them. The digital image is an indispensable tool for fieldworkers. This chapter comprises of a case study that illustrates how digital photography can be employed to establish a digital archaeological record based on photographs taken in the field of headstones located in several urban churchyards situated within central Oxford.

Keywords: Adobe photoshop, digital cameras, images, pictorial representation, repeat photography/rephotography, quantification, visual records.

Since the time that Louis Daguerte captured an image using the camera obscura in Paris on the 19th of August 1839 (Langford, 1978), photography has come a long way. With a vast potential to capture (freeze) visual reality, photography could be employed as optical instruments in scientific research. They could be deployed, for instance, to capture details at various spatial and temporal scales. Microscopes and other imaging devices could provide close-up images that could be captured photographically at several thousand times the normal size, as in the case of the scanning electronic microscope or SEM. What is more is that digital cameras do not depend on processed film for the creation of photographs and, thus, have allowed for the application of photography simply and inexpensively indoors and outdoors as a portable optical recording device.

Because it is regarded as truthful and accurate, photography can be trusted to provide material (recorded moments in time) for comparative analysis. Their realism and detail allow for cross-temporal comparisons, for instance, drawn from historical records. Before 1839, in fact, visual records were limited to drawings, paintings, and engravings. Since the 1870s and 1880s in particular, hand cameras and faster and more sensitive photographic materials enabled instantaneous photographs (Langford, 1978). This was crucial for the application of photography outside the arts, including in scientific research. Before-and-after comparisons could be made, recording change across time. Contours of change could be derived from such photography quantitatively. Indeed, before digital

Mary J. Thornbush and Sylvia E. Thornbush

photography, photographic film and prints had to be scanned in order to derive an image for manipulation. This introduced error into quantitative applications.

Since digital photography does not require the processing of film, the error associated with this was eliminated and digital quantification improved. As a tool, quantitative digital photography could be employed repeatedly as part of the rephotography methodology to measure landscape change (Webb, 1996; Webb *et al.*, 2010) and environmental history (*e.g.*, Thornbush, 2008a). In this way entire expeditions can be revisited and modern photographs taken for comparison with the historical record (*e.g.*, Webb, 1996). Several works by Thornbush (*e.g.*, 2010b) have employed digital rephotography in order to qualitatively assess changes to soiling and decay of the limestone ashlar on building exteriors in central Oxford. This was performed through the simple comparison of before-and-after images based on some archival photographs as well as digital photographs taken in photographic surveys. Where greyscales were included, however, this constituted a more quantitative approach that will be covered in the next chapter on quantitative photography. She also employed this methodology (of digital rephotography) to track the environmental history of climbing plants in central Oxford (*e.g.*, Thornbush, 2008a, 2013c).

Digital photographs could also be employed in a more straightforward fashion as a pictorial instrument (Swallow *et al.*, 2004), recording objects in time and space. In this way, photo-pairs could be compared and change assessed through the differences depicted in the before-and-after images. This methodology could be employed to capture both temporal and spatial differences. Singly, images could be taken for various purposes, including to illustrate something in the field. Thornbush (2012), for instance, used close-up digital photography in order to capture decay features found along a laneway in central Oxford. She was able to employ these images to depict weathering forms in her index, namely the size-extent (S-E) index. This weathering scale was later applied to headstone weathering to churchyards located in central Oxford (Thornbush & Thornbush, 2013).

Human geographers have also employed photographic techniques in their research (Rose, 2008), and (disposable) cameras were worked into undergraduate fieldwork as a cheap recording device (Sidaway, 2002). This approach could provide a basis for establishing a web-based documentary of fieldwork, improving the sharing of the experience and also acting as a real-time record (when uploaded to course webpages and social media accessible *via* the Internet. Photographs (slides) are already considered to be a classic (established) tool to be used as a passive learning approach (Jones, 2000). However, with the mass production of digital cameras, they became less expensive to purchase and incorporate into research studies and field-based instruction.

The digitisation of photographs has enabled for online records and digital databases, as for instance ViewFinder, available from English Heritage (2011), as a digital database of historical photographs that can be accessed online. Other collections, including archival and museum collections, have been digitised and linked to the Internet. For instance, photographs taken by Harry Burton of the Tutankhamun archaeological expedition are available online through the Griffith Institute (2005). The V&A (2014a) is another example of digital collections now available online. The Ashmolean Museum (2012) website also houses several online collections, including aerial photographs taken by Major George Allen (1891-1940) and photographic archives. It is becoming increasingly popular to establish digital libraries for collections, making them more easily and widely accessible.

In this way, photographic collections could be examined virtually, allowing for improved access and expanding the possibilities for new desk-based studies. With fewer constraints on access, any scholar from around the world is now able to examine digital picture libraries and archives as well as develop research based on these collections.

Digital photographs have also facilitated the recording of experiments and the results of scientific research, improving the sharing of information and its availability to other researchers and scientists. Microscopes now have the capacity to capture digital images, and these can be made available through published micro pictograms in figures. The technology has developed to include field-portable instruments, such as the USB digital microscope (*e.g.*, PCE-MM200), that has the capacity to take digital images with up to 200 times magnification.

Having an image available is crucial to demonstrating evidence of something or for verification purposes. For instance, it is possible to easily and quickly send an image as an email attachment to a colleague in order to verify a type of microbe based on a microscopic image. So, this has potential to improve the quality of publications and contribute towards improvement in science, including interdisciplinary research. Having photographic evidence at different scales opens up new possibilities and allows for cross-disciplinary collaboration in unravelling a problem.

Most importantly, however, is the use of digital photographs as a basis for establishing an electronic record that compliments any print documents. Not only is an electronic copy easy to disseminate, it is also possible to establish a virtual record as part of an online database or digital collection. This is indispensable for historians, archaeologists, and others who work with historical objects or documents. The process of establishing and developing a virtual record that is available online is a way of maintaining the integrity of collections, as they themselves age, and to allow for cross-temporal comparison through salvaged collections.

The subsequent case study employed a field-lens method to visualise biological growths on headstones in Oxford churchyards. This was coined the 'photoscope technique' and was deployed for fieldwork before the knowledge (by the authors) of any portable microscopes. However, there have since been such developments, and portable USB microscopes are now available, simplifying the process of close-up visualisation and image capture in the field.

Case Study: Weathering Limitations of Headstone Seriation since the 17th Century at Four Churchyards in Central Oxford, UK

Abstract: In establishing a seriation of headstones, it is necessary to consider the integrity of the available record and, hence, the effects of weathering on preservation. Headstone shape, motifs, and introductions were surveyed at four centrally located churchyards in Oxford, including St Giles, St Mary Magdalen, St Peter-in-the-East, and St Cross. A 'photoscope technique' was incorporated to magnify, directly in the field, the type of weathering damage on limestone headstones. Several details were derived from this study to include headstone shape (slightly curved), the most commonly used motif (cross), and the typically used introduction ('Sacred to the memory of'). Of the legibility, slightly less than one-third of headstones had legible dates. Chemical dissolution, pitting, and encrustation were frequently observed with some physical weathering, such as sheet exfoliation, and biological weathering.

INTRODUCTION

The examination of headstone seriation is important in classifying changing trends in burial patterns found in churchyards. This type of study is not new to archaeology, and was investigated by Dethlefsen and Deetz (1966) in a classic seriation of headstone motifs and epitaphs in eastern Massachusetts, USA for the 17th and 18th centuries, where they also aimed to decipher religious change in different geographical regions. Important studies in the field of headstone iconography include research conducted in Scottish graveyards, as with Willsher's (1985) *Understanding Scottish graveyards*, which examined the different motifs, lettering, and epitaphs on headstones. Further publications on inscriptions from gravestones in Scotland date back to the early 18th century (Ferguson, 2002, p. 8). For England, Mytum's (2000) *Recording and analysing graveyards* examined similar features to Willsher's (1985) study, focusing on the elaborateness of headstone shape. For Ireland, Longfield (1946, 1954, 1955) published extensively on 18th century Irish headstones, addressing varying quantity and quality of headstone motifs as well as the loss of such motifs through rock weathering. For graveyard management, Strangstad's (1988) *A graveyard preservation primer* examined the replacement and maintenance of graveyards in the USA. Other research relating to tombstones includes Keister's (2004) *Stories in stone: A field guide to cemetery symbolism and iconography*, which addressed cemetery symbolism and iconography in various locations around the world. Clairmont's (1993) *Classical Attic tombstones* categorized markers by the number of the interred in each plot, and is extremely relevant for future demographic studies.

The weathering of headstones appears, for instance, where there are surface dissolution features (Hoke & Turcotte, 2004). Atmospheric acidity is especially pronounced in urban areas, where there is enhanced weathering, as found by McNeill (1999) for Scottish gravestones, Cooke *et al.* (1995) for Swansea and other sites in England, and Inkpen and Jackson (2000) for southern Britain. Surface reduction has even been observed for hard stone, such as marble monuments, linked to industry in North American and European cities (Dragovich, 1987).

The current study includes an examination of stone decay on the archaeological record in centrally located Oxford churchyards. The use of churchyards in this study stems from the small range of spatial variation available in these, providing an opportunity to examine 17th to 20th century dynamics of commemorated burial (Mytum, 2000). The sites considered here are: 1) St Giles churchyard (SGC) located in the northern part of central Oxford, 2) St Mary Magdalen churchyard (SMMC) centrally situated in central Oxford, 3) St Peter-in-the-East churchyard (SPEC), and St Cross churchyard (SCC), both found in the eastern part of central Oxford (Fig. **1**). Headstone shape, motifs, and introductions were examined and classified in order to construct a seriation. In addition, different weathering types (chemical, physical, and biological) and respective features on the headstones were investigated, and the extent to which weathered details hindered a complete seriation is discussed.

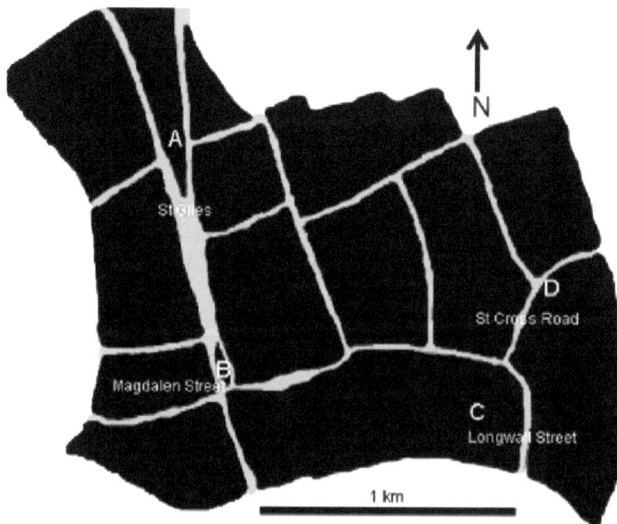

Figure 1: The positioning of each churchyard within the Oxford city centre: a) St Giles, b) St Mary Magdalen, c) St Peter-in-the-East, and d) St Cross.

The current study area in the northern, central, and eastern parts of central Oxford shares a similar temporal period of church construction. The churches in all four churchyards were constructed in and around the 12th century. The churchyard belonging to St Giles was constructed outside the main town centre in the 12th century and a large array of people were interred, among them the rich, the poor, servants, and the clinically insane. By the 19th century, it was disused due to population growth, terminating any continuum of headstone styles in this churchyard. St Mary Magdalen was the most densely populated church in the 16th century, at a time when the church served all social classes, such as gentlemen, college servants, crafts workers, labourers, and the poor. However, by the 18th century, there was a marked social division, and by the end of the century it became a conventional Anglican church. St Peter-in-the-East, which is located at St Edmund Hall, University of Oxford, is a Norman church now used as a library, being active as a church from 1559 to 1965. The interred in this churchyard range from former University staff to common people. St Cross church was active in and around the 16th century. Of the interred, there were a good number of high social status individuals between 1550 and 1850, but the churchyard was later disused in 1855. The register available contains burials in the 17th century, but any subsequent records were ill-managed and, thus, not recorded effectively (Jones, 2009).

MATERIALS AND METHODS

Permission was acquired verbally and in advance for all four sites. Each churchyard was mapped (maps for each site were unavailable) and historical information was gathered. Sites were photographed to assist with mapping, as well as to retain a digital record of headstones, which could be useful for future use (*e.g.* assessments). At each site, each headstone was mapped and then photographed at approximately a meter's distance, with a Nikon Coolpix S4 digital camera. Dimensions of height, width, and thickness were measured with a measuring tape, and these data were recorded in tabular format. The date, headstone shape, motifs, introductions, names of the deceased, and epitaphs were all recorded, where available. Headstone characteristics were classified according to Mytum (2000). Visual assessments were made of weathering based on a 'photoscope technique' adopted for field-use of headstone close-ups, using a hand-held lens (with a magnification of 30) in conjunction with the digital camera, as demonstrated in Fig. **2**.

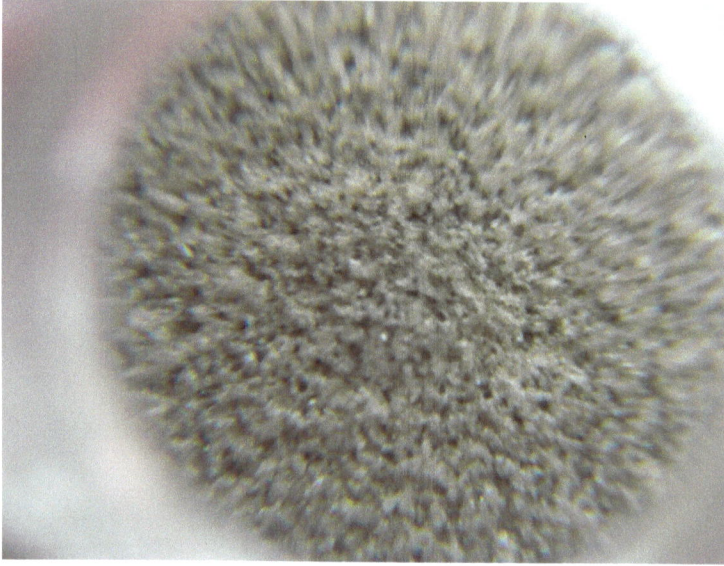

Figure 2: Photograph taken at St Giles, demonstrating the field-based 'photoscope technique'.

RESULTS

The two most common headstone shapes photographed in the four Oxford churchyards are slightly curved (4500) (Fig. **3a**) and triangular (4300) (Fig. **3b**), and these varieties are most commonly used in the first half of the 19th century (Table **1**). Less commonly used varieties of headstones photographed in the four churchyards are flat (4700) (Fig. **4a**), sinuous (4600) (Fig. **4b**), semicircular (5100) (Fig. **4c**), round (4100) (Fig. **4d**), Gothic pointed (4200) (Fig. **4e**), and slightly-curved-flat-shouldered (5300) (Fig. **4f**). All but one of these headstone shapes date back to the 19th century, with the majority dominating the first half of the century, as shown in Table **1**.

Figure 3a: Photograph taken at St Giles, demonstrating the headstone shape 4500 (slightly curved) and the crown motif.

Figure 3b: Photograph taken at St Peter-in-the-East, demonstrating the headstone shape 4300 (triangular).

Figure 4a: Photograph taken at St Peter-in-the-East, demonstrating the headstone shape 4700 (flat).

Figure 4b: Photograph taken at St Mary Magdalen, demonstrating the headstone shape 4600 (sinuous).

Figure 4c: Photograph taken at St Giles, demonstrating the headstone shape 5100 (semi-circular) and the cross motif.

Figure 4d: Photograph taken at St Peter-in-the-East, demonstrating the headstone shape 4100 (round).

Figure 4e: Photograph taken at St Giles, demonstrating the headstone shape 4200 (Gothic pointed).

Figure 4f: Photograph taken at St Mary Magdalen, demonstrating the headstone shape 5300 (slightly-curved-flat-shouldered).

Table 1: The dominant and less commonly found headstone shapes at four central Oxford churchyards.

	Dominant		Less Common					
	HS4500	*HS4300*	*HS4700*	*HS4600*	*HS5100*	*HS4110*	*HS4200*	*HS5300*
1650-1699	1			1				
1700-1749	1	1			1			
1750-1799	2		1		4			
1800-1849	38	15	13	9	7	3	2	1
1850-1899	12	7	2	4		1	2	
1900-1949	1			1	2			
1950-1999	1							
Total	56	23	16	15	14	4	4	1

The majority of motifs are found on headstones pertaining to St Giles; however, the proportion of headstones containing motifs is small (Table **2**). The most commonly used designs are the cross motif (see Figs **3.3b** and **3.4c**) and floral motifs (Fig. **5a**), and, to a lesser extent, the scroll (including the scroll pattern) (in Fig. **5a**), crown (see Figs. **3a** and **5b**), and IHS (in Fig. **5b**) motifs. Less frequently used designs include death's head (Fig. **5c**), (sun with) radiance (Fig. **5d**), and chalice (Fig. **5e**) motifs. Most motifs appear in the 19th century, with the majority dominating the first half of the century (Table **2**).

Figure 5a: Photograph taken at St Giles, demonstrating floral and scroll motifs.

Figure 5b: Photograph taken at St Mary Magdalen, demonstrating crown and IHS motifs.

Figure 5c: Photograph taken at St Peter-in-the-East, demonstrating the death's head motif.

Figure 5d: Photograph taken at St Mary Magdalen, demonstrating the (sun with) radiance motif.

Figure 5e: Photograph taken at St Giles, demonstrating the chalice motif.

Table 2: Motif types found at four central Oxford churchyards

	Cross	Floral	Crown	IHS	Scroll	Death's head	(Sun with) radiance	Chalice
1650-1699					1	1		
1700-1749								
1750-1799	1							
1800-1849	4	5	3	1	2		2	1
1850-1899	3	1	1	2				
1900-1949	1							
1950-1999								
Total	9	6	4	3	3	1	2	1
SCC	2	2					1	
SGC	3	2	3	1	2			1
SMMC	3	2	1	2			1	
SPEC	1				1	1		

Of the introductions, the most popular introductions are 'Sacred to the memory of' and 'In memory of', which appear mostly in the first half of the 19th century and to a lesser extent in the second half of the century (Table **3**). All other introductions, however, do not seriate because the sample is too small. The appearance of most introductions occurs at St Giles, with only a few headstones being devoid of these introductions (see Table **3**).

Table 3: Introductions found at four central Oxford churchyards

	Sacred...	*In memory...*	*Here...*	*To the memory...*
1650-1699			1	
1700-1749		1		
1750-1799	1	3		
1800-1849	63	15	2	5
1850-1899	15	4		
1900-1949	1	1	1	
1950-1999		1		
Total	80	25	10	5
SCC	27	12	1	4
SGC	25	8	2	1
SMMC	21	4		
SPEC	7	1	1	
	In piam...	*Beneath...*	*Erected...*	*Total HS*
1650-1699				1
1700-1749				1
1750-1799				4
1800-1849				85
1850-1899		1		20
1900-1949	1		1	5
1950-1999				1
Total	1	1	1	117
SCC				44
SGC	1			37
SMMC		1		26
SPEC			1	10

The illegibility of headstones in this study is a great limitation, as a large proportion of the headstones photographed are severely weathered. The majority of weathered headstones appear at St Peter-in-the-East; and the headstones with legible dates are mostly found at St Cross (Table **4**).

Table 4: Legibility of headstones at four central Oxford churchyards

	SCC	*SGC*	*SMMC*	*SPEC*	*Total HS*	*Total (%)*
Legible	52	46	28	11	85	0.29
Weathered	3	50	79	78	207	0.71
Total HS	55	96	107	89	292	
Legible (%)	0.95	0.48	0.26	0.12		
Weathered (%)	0.05	0.52	0.74	0.88		

Chemical weathering, especially dissolution or carbonation and encrustation, is the most prominent type of weathering damage evident at the churchyards examined in this study. Since almost all headstones were comprised of limestone, they are susceptible to decay from acidity that varies from natural levels of gases, such as carbon dioxide (CO_2), occurring in rainfall that enhances rainfall acidity from combustion emissions, such as from vehicular exhaust. This acidification of rainfall in urban areas causes much dissolution of limestone, leading to reduced detail in inscriptions and motifs carved into the headstones. Fig. **6** from St Giles conveys an example of dissolved details in the lettering of the headstone belonging to William Taylor, which has an indiscernible date and was, thus, excluded from the seriation.

A common chemical weathering feature associated with limestone dissolution is pitting, and it can be found on various headstones, as evident in Fig. **7a** from St Giles, where this dissolution feature has erased all details. Alveolar and cavernous weathering are evident on a headstone of the Gothic style with extensive damage due to dissolution that is moving inwards on the inscribed surface (Fig. **7b**). Chemical dissolution from dissolved salts in groundwater rising to the surface (due to capillary rise) can lead to near-ground deterioration, as evident in Fig. **7c** at St Peter-in-the-East. Encrustation is visible through the development of black crusts, as evident in Fig. **8** at St Mary Magdalen. The formation of crusts results from the accumulation of deterioration products on limestone, such as gypsum, which darken through exposure in polluted environments, where the atmosphere is loaded with particulate (Fassina *et al.*, 2002). Once crusts are breached, they expose friable unconsolidated material beneath, as in Fig. **4e** at St Mary Magdalen, which in this case is exfoliating towards the bottom of the marker, indicating some impact of capillary rise.

Figure 6: Photograph taken at St Peter-in-the-East, demonstrating dissolved details of inscriptions on a limestone headstone.

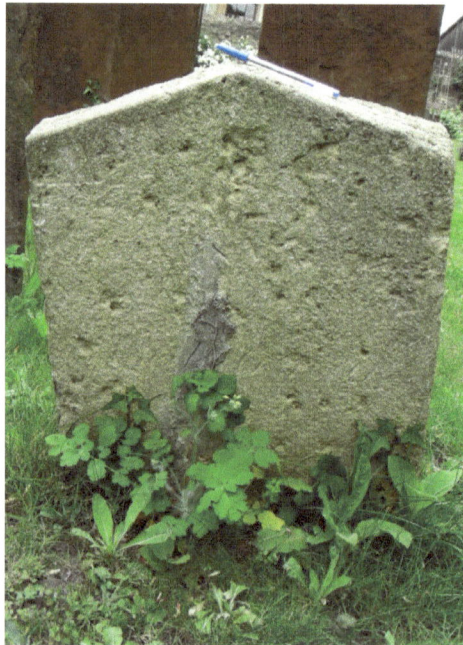

Figure 7a: Photograph taken at St Giles, demonstrating pitting of limestone.

Figure 7b: Photograph taken at St Mary Magdalen, demonstrating alveolar and cavernous weathering of limestone.

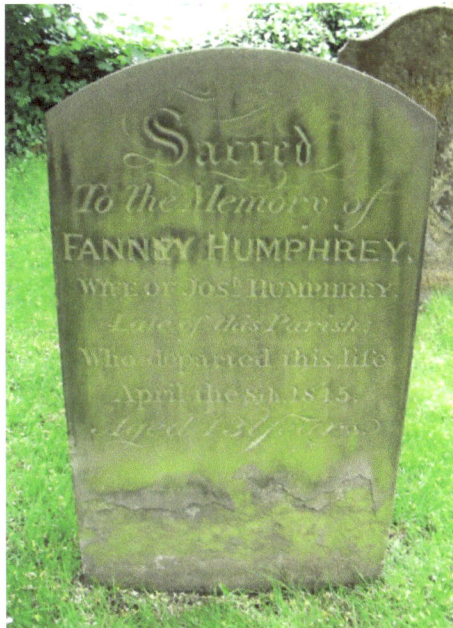

Figure 7c: Photograph taken at St Giles, demonstrating the effect of capillary rise on a limestone headstone.

Figure 8: Photograph taken at St Mary Magdalen, demonstrating encrustation (black crust) on limestone.

In terms of mechanical or physical weathering, there is evidence of mainly flaking and exfoliation. Flaking, as seen in Fig. **9a** at St Giles, occurs when pieces on the surface of the headstone crack, fracture, and peel off due to expansion-contraction of the material (due to heating and cooling) when exposed to temperature fluctuations in an outdoor environment. The 'photoscope technique' enabled a close-up view of flaking, which appears in Fig. **9b** at St Giles. The process of expansion-contraction from heating-cooling episodes can also lead to the exfoliation of bedded material, such as limestone, where there is a relative weakness in the stone. Sheet exfoliation of these limestone headstones is apparent in particular, at St Mary Magdalen, as evident in Fig. **9c**. This type of weathering detaches layers of material, often on which there is information inscribed, including the date. Exfoliation was focused towards ground-level in some cases, such as the headstone from St Mary Magdalen appearing in Fig. **5d**.

Figure 9a: Photograph taken at St Giles, demonstrating (close-up) flaking of limestone.

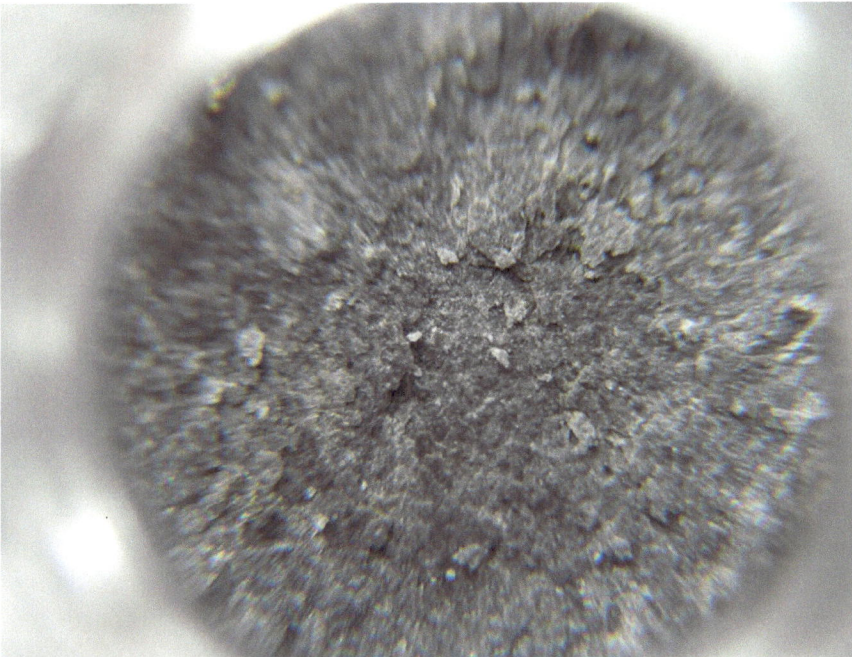

Figure 9b: Photograph taken at St Giles, demonstrating flaking of limestone using the 'photoscope technique'.

Figure 9c: Photograph taken at St Mary Magdalen, demonstrating sheet exfoliation of limestone.

There was considerable biological weathering at these sites, both in terms of flora and fauna at various scales. An algal residue was evident at most locations, which was at times more developed underneath trees. This residue often develops in moist (wetted) areas, such as along north-facing streets in Oxford (Arkell, 1947). There was an abundance of lichens found in these churchyards. The headstone most colonised by lichens appears in Fig. **10a** at St Giles. A close-up of lichen colonisation can be seen in Fig. **10b**. Moss can be seen at St Giles, as in Fig. **10c**, and is abundantly apparent in Fig. **10d** at St Mary Magdalen. Moss grows on headstones in wetted areas, where snails can also be seen (Fig. **10e**) at St Giles. Also here, insect colonisation is evident and, in particular, egg-laying on headstones (Fig. **10f**). Finally, at St Peter-in-the-East, English ivy (*Hedera helix*) can be seen growing onto headstones (Fig. **10g**), along with a profusion of other vegetation (see Fig. **5c**).

Figure 10a: Photograph taken at St Giles, demonstrating the colonisation of a limestone headstone by lichens.

Figure 10b: Photograph taken at St Giles, demonstrating a close-up of lichen colonisation on limestone.

Figure 10c: Photograph taken at St Giles, demonstrating the growth of moss on a limestone headstone.

Figure 10d: Photograph taken at St Mary Magdalen, demonstrating an abundance of moss growth on a limestone headstone.

Figure 10e: Photograph taken at St Giles, demonstrating snails on limestone near the appearance of moss growth.

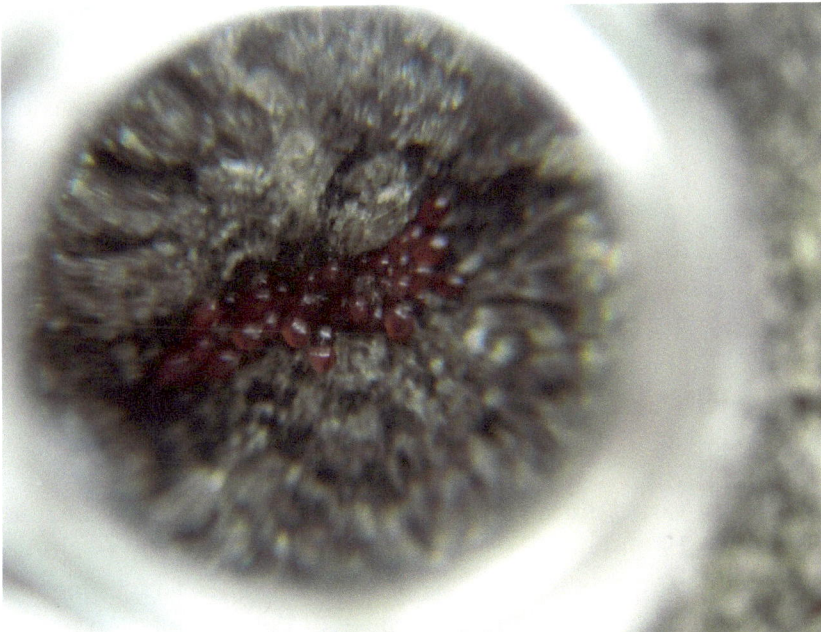

Figure 10f: Photograph taken at St Giles, demonstrating a close-up of insect eggs lain on a limestone headstone.

Figure 10g: Photograph taken at St Peter-in-the-East, demonstrating English ivy (*Hedera helix*) growth on a limestone headstone.

DISCUSSION

The types of headstones photographed and documented at the four churchyards are influenced by their history, such as church disuse or social division. When St Giles was disused in the 19th century, due to rapid population growth in the area, people were later interred north of this location, minimising any new styles in the churchyard. St Mary Magdalen, on the other hand, underwent a marked social division by the 18th century, which in several instances had a profound influence on all aspects of headstone style primarily due to a monetary influence (Mould & Loewe, 2006), where simplicity dominated the overall appearance of the headstones. St Peter-in-the-East continued to inter its dead into the 20th century, though this century marked the start of a more simplistic headstone style, unlike its predecessors in the 19th century (Mytum, 2000).

Headstone shapes varied throughout each 50-year period (used in this study) between 1650 and 1999; but, of the dominant shapes, the 4500 (slightly curved, see Fig. **3a**) variety appeared throughout the time-line, suggesting that these headstones were mass-produced either because of their easiness to carve or their popularity from the 17th century through to the 20th century in these Oxford churchyards. As with the popularity of other headstone shapes, these appear throughout the entirety of the 19th century. This clustering of headstones is attributed to either disuse or social division within these churchyards.

Of these styles, it is unknown whether a headstone design was chosen by the deceased or their family; however, irrespective of who did the selection, influential factors such as economics and religion played an important role in types of headstone shape, motifs and inscriptions used. Nevertheless, in terms of headstone shape, other issues could also have influenced the type. For instance, the largely used 4500 headstone variety could also be attributed to mass production by stonemasons. Mass producing a headstone shape style saves the stonemason time, especially during a period of time where the death rate is high (Walker, 1983), but not to the point where demand of headstones exceeds the supply rate, leading to the absence of a marker. Any other style or alteration of this shape would then have been more expensive and, thus, would have been limited to those families who could afford them. The costs incurred from carving an entirely new headstone would have come from the family, not to mention the work required by the stonemason and carver. However, any scarcity of stonemasons and carvers in the area could also be attributed to this simplistic style (Gorman & DiBlasi, 1981).

The limited use of motifs can be attributed either to the weathered state of the headstones, preventing any recording of motif and its respective date, or to social and economic status, playing an influential role. Post-1900 headstones are simplistic in style, unlike headstones dating from the Victorian period. Some studies have looked at the differing styles of similar motifs with a focus on carvers, but this was not an aspect of this study. For instance, Fraser and Rönne (1957) described scroll motif elaborateness in two classes, namely flower-scrolls and spiral-scrolls, where the former is subdivided into three categories: wave-scroll, vine-scroll, and pediment-scroll. It is this distinctiveness that must be preserved to prevent any further destruction of detail (Mould & Loewe, 2006). Similarly, Longfield (1946, 1954, 1955) looked at motifs on Irish headstones, exclusively at the changing styles of biblical scenes, giving importance to the carver's role.

Headstone motifs are influenced by social, economic, and religious factors (Dethlefsen & Deetz, 1966; Gorman & DiBlasi, 1981). The use of motifs takes on a dynamic function of social class and religious values (Gorman & DiBlasi, 1981; Keister, 2004). Mytum (2000) and Keister (2004) described the various styles represented on headstones throughout time, such as Baroque and Rococo floral features, Gothic revival features, such as the cross and IHS in various styles, and

Anglican revival scroll decoration that were present at the four churchyards studied, while Keister (2004) presented an elaborate description of cemetery funerary architecture, including examples of the above from around the world.

Of the influential factors, religion appears to be the most significant, with the popularity of crosses on headstones. Floral motifs were especially popular in the mid-19th and 20th centuries as symbols of remembrance (Mytum, 2000), and are secondly ranked in terms of popularity. However, the connection between simplicity and economics cannot be fully established here as, though headstones that lack motifs contained dates, others without dates contained motifs that were rarer in these churchyards.

Introductions used are 'Sacred to the memory of' and 'In memory of', which are influenced by a salvation theme, where sadness is masked with the idea that the soul of the interred has found peace (Mytum, 2000). In contrast, the mortality themes, such as 'Under this monitor' or 'Here lies the body of', state the grim truth about the deceased with no air of hope for what the living descendants have lost (Mytum, 2000). The pattern of appearance of introductions was significantly clustered at St Cross and St Giles. Nevertheless, all churchyards except for St Peter-in-the-East contained a similar pattern of popularity, a trend that began to thin out after the mid-19th century (see Table **3**).

Overall, the reason for the small sample size is primarily due to the weathered state of most of the headstones photographed. Even though the shapes or motifs are evident, without dates these cannot be used in the seriation. There are many factors that weather or erode the surfaces of headstones. Paine (1992) described several factors that affect the appearance of headstone surfaces, including the use of herbicides and pesticides, relocation of headstones, scrape marks made by maintenance equipment, and vandalism. Herbicides and pesticides can damage headstone surfaces through chemical reactions (Fraser, 2002), moving headstones and the use of machinery (*e.g.* lawn mowers) are also culprits in the destruction of these (Paine, 1992). At St Peter-in-the-East, relocation of headstones influenced, to a degree, the destruction of these (and as a current map was unavailable, motifs on dateless headstones were unrecorded). Consequently, the lack of recording unprecedented motifs that appear on weathered headstones affects this type of study. However, a growing concern of headstone destruction is due to vandalism, whether direct or indirect. Walking away with parts of headstones (Foster, 2002) or leaning up against them while engaged in leisure activities represent different degrees of vandalism. All four churchyards in Oxford contained tilted or fallen headstones, and some had microbial growths particularly in the presence of trees or large shrubs, as found in old public cemeteries in the City of Providence, Rhode Island, USA (McMahon, 2002). Buckham (2002), who is Carved Stone Advisor for the Council for Scottish Archaeology, views gravestones as resources that are alarmingly deteriorating, and examined the recording (stone type, design, and decoration), management, and policy of Scottish graveyards.

Chemical dissolution from high levels of pollution in urban areas as well as mechanical weathering along points of weakness in the rock fabric, in conjunction with biological incursions of algae, lichens, and insects, contribute to a lack of integrity in the record. However, by examining seriations it is possible to decipher something of the temporal framework of headstone types; for example, this study has shown that 4500 types were used in Oxford predominantly in the 19th century. In places like Scotland, for instance, most headstones date no earlier than the 17th century (Urquhart, 2002), and of the headstones dating to this period, in the current study, only one contains a date, yet some dateless headstones contained motifs that were observed to be from the same century, emphasising the severe loss of data in these Oxford churchyards. As stated by Rodwell (1981), the recording and photography of headstones prior to the 20th century is important for current research. Regardless of weathering, there are unique features in the three examined churchyards, for example, St Peter-in-the-East contains a record of the death's head motif (Fig. 5c). Even though some details have been weathered, such as inscriptions, with dates, some unique features are still evident in these churchyards.

The need for education of headstone preservation is growing and has been carefully planned, as by Historic Scotland. Similarly, Foster (2002) and Ferguson (2002) have been involved in innovative methods of preservation from new policies to web-accessible databases and the use of global positioning systems (GPS) in graveyard reconstruction.

CONCLUSION

Headstones are rapidly becoming a thing of the past, as cremation practices are becoming increasingly popular, yet their documentation and preservation are slow in progress. With the advent of limited burial spaces funerary rituals are quickly changing due to spatial preservation. The growing concern for the preservation of headstones is hindered by their continued indirect and direct destruction. The need for further and similar research is important to not only understand the changing patterns of past societies, but also to appreciate the uniqueness of the styles utilised in different societies in history.

Fraser (2002) introduced preservation programs directed at improving the overall appearance of headstones and monuments in graveyards, such as 1) guided tours, audio-guided tours, and displays highlighting the significance of headstones in graveyards, 2) seminars specifically related to graveyard preservation hosted by groups specialising in graveyard history, 3) brochures explaining the conservation methods in cleaning headstones and monuments, and 4) staff seminars where departmental research is applied, if applicable, to headstone and monument preservation.

Other events, such as some ghost tours, are more indirectly connected to graveyard preservation. For example, Edinburgh's City of the Dead ghost tour is centred inside Greyfriars kirkyard, which exerts emphasis on the overall upkeep of the kirkyard, leading to the management of important headstones connected to the popularity of the ghost tour (Black Hart Entertainment, 2010). Essentially, whilst public interest remains, there will be headstone preservation.

ACKNOWLEDGEMENTS

The authors are grateful to the porters at St Edmund Hall, University of Oxford and the staff at St Giles and St Mary Magdalen churches for their assistance and cooperation. Acknowledgement also goes out to Dr John Jones and Mr Alan Simpson for their assistance in providing a brief history of St Cross church and churchyard.

CHAPTER 4

Quantitative Photography

Abstract: A recent, and what seems like a natural evolution, of photography has been for quantification. With modern cameras that can more realistically depict what is visible, it is possible to measure change using (digital) photography. This chapter considers various examples of the application of photography in the measurements of surface colouration (soiling) and the decay of weathering forms. Techniques developed by the first author are presented, and a case study on the application of the O-IDIP method is presented to convey photographic surveys for the quantification of stone surface colouration associated with biological colonisation on a string course in central Oxford.

Keywords: Calibration, CIE *Lab*, decay mapping, DMAP approach, histogram-based colour measurement, O-IDIP method, surface colouration.

The first-listed author first became involved in this area of research during a photographic survey in central Oxford during her doctoral studies. A stranger approached her research team and casually asked what they were doing. Her former mentor responded by explaining that they were conducting a photographic survey of rock weathering. His response was that there was no real point to a subjective approach that could not be used to measure anything. This interaction sincerely intrigued the author, who began to brainstorm about how to make photographic surveying more objective and quantitative. This approach has been developed since 2004 and continues into the present day. This chapter addresses her contribution to the field and provides an overview of other published works.

It is possible to make photography a more rigorous measurement methodology. This is evident through the integrated digital photography and image processing (IDIP) method devised by Thornbush and Viles (2004a), which has been subsequently broadly applied in central Oxford. Specifically, digital photographs are taken mounted on a tripod for a consistent elevation perspective of images. Ideally, a digital camera is employed in order to reduce any potential error associated with processing film and scanning in image capturing and acquisition. Moreover, a similar distance is maintained throughout the photographic survey, which is usually performed close-up to the stonework with a metre or so of walls. Images are compared either cross-spatially or -temporally in order to assess any changes. For instance, Thornbush and Viles (2004b) applied the IDIP method to Bath limestone sensors that had been exposed between 1996 and 2006 at roadside locations in busy as well as quieter streets in central Oxford. Using CIE *Lab* in

Mary J. Thornbush and Sylvia E. Thornbush

image processing, they were able to capture darkening of these stone surfaces after exposure, including a patchy distribution of soiling. More specifically, a spotted patchiness was evident after just a year of exposure and was the dominant soiling pattern at the sensor scale on the longest exposed disc sensors.

Various aspects of this method make it quantitative. For one, when a colour chart is incorporated into photographs it is possible to calibrate image brightness and colouration. This especially becomes an issue with outdoor photography. The initial testing of this method (Thornbush & Viles, 2004a) was performed in an indoor setting inside a laboratory at the School of Geography and the Environment. Their second paper (Thornbush & Viles, 2004b) was also based on digital images acquired indoors. However, for an environmental application, outdoor settings are more realistic and the need to develop such a modified approach became apparent. Other authors, such as Searle (2001) incorporated a greyscale into their outdoor photographs of building ashlar in order to depict the soiling of Bath stone. Similarly, Viles had already incorporated a plastic-covered greyscale into previous photographic surveys before Thornbush began her doctoral work. In her repeat photographic surveys, she used an exposed paper greyscale in order to reduce glare from a plastic-covered greyscale. Various papers were published of photographic surveys of walls derived from this integration of a greyscale, comparing cross-temporal and spatially variant locations in the city centre. For instance, Thornbush and Viles (2008) considered photographs from 1997 in comparison to those taken in 1999 and 2003, with images taken consistently at 30 × 30 cm in dimensions at 1-1.3 m above pavement level. Images captured minor changes, but quantitative changes were possible to access through pixel-based size proportions of specific decay features (blistering, flaking, and scaling). This paper was based on greyscale images and found that colour images are more informative about soiling and decay. It also indicated the problems associated with consistent outdoor lighting conditions affecting the results.

Thornbush (2008b) independently published a technique for the quantification of colour images. This was based on three main channels: black-white (L), green-red (b), and blue-yellow (a). Images were taken in 2006 on two different days, of clear sky *versus* overcast conditions. A spectrophotometer was used for validation of outdoor lighting conditions across photographic surveys. It was found that simple greyscale calibration improves the comparability of surveys to spectrophotometric measurements. On the basis of these results, the outdoor application of the IDIP, known as the O-IDIP, method was developed. For instance, Thornbush (2010a) published a case study based at the Ashmolean

Museum along its southern façade. In this work, she continued to use histogram-based quantification of camera-captured images to track, this time, surface colour change between 2005 and 2007, again, in different outdoor lighting conditions of a clear sky *versus* overcast. Calibration in this study was based on spectrophotometric data collected in the winter of 2006. It supported that the calibrated O-IDIP method was able to measure change before *versus* after the cleaning of the southern façade, which was darker in 2005 (with a lower level of lightness), especially at the east elevation. It revealed that lightness (surface darkening or blackening) is more affected by outdoor lighting conditions than chromatic values along green-red (*a*) and blue-yellow (*b*) channels of colour, meaning that there is more error associated with soiling measurements at the façade-scale (in an outdoor setting). This means that scale is an issue that should be considered in the mapping of decay features, as considered by other authors, such as Inkpen *et al.* (2001). Inkpen *et al.* (2008) subsequently employed GIS in order to examine stone degradation. In comparison, the O-IDIP allows for calibrated outdoor photography and also provides a simple approach to quantification using digital images.

More recent work has applied the O-IDIP method to measuring urban greening. One such application has been to quantify algae growing on north-facing walls in central Oxford (Thornbush, 2013b). This allowed for the detection of moss growth with surface depressions on building ashlar through increases in standard deviation (Std Dev) values for north-facing walls. The method was also able to detect changes associated with rainwashing; specifically, rainstreaks were measured through the *a* channel. Most recently, Std Dev values have been used to quantify surface roughening associated with decay (Thornbush, 2014b). Taking digital photographs under conditions of direct sunlight (clear sky) *versus* indirect sunlight (overcast), it was possible to derive a quantification measure of surface roughness through the Std Dev of *L* of calibrated images. The resulting roughness index is low when surface roughness is low (cast shadows are limited) and increases with the casting of more shadows under conditions of direct illumination by sunlight in comparison with indirect radiation. Precipitates are known to increase surface roughness as they become weathered. In this project, she was also able to capture evidence of biopitting by climbing plants at the University of Oxford Botanic Garden; however, she discovered that biopitting does not increase measured surface roughness using this application of the O-IDIP method.

In addition to the IDIP method and its outdoor derivative as the O-IDIP method, other approaches have been devised to measure decay features on walls. For instance, Thornbush and Viles (2007a) examined the boundary wall of Worcester College, Oxford by employing a novel approach, namely the decay mapping in Adobe Photoshop (DMAP) approach. This study measured changes in surface brightness and roughness based on close-up photographic images of walls, applying the Magic Wand Tool to greyscale images in Lab Color Mode to select proportions of pixels with a lightness (*L*) value of 77%. The DMAP approach proved to be particularly useful when applied to long-term monitoring exceeding 5 years of survey.

The final case study in this part of the eBook uses the O-IDIP method to quantify the appearance of lichens colonising a string course in front of the University Church of St Mary the Virgin in central Oxford. This building and structure are located close by a main roadway (High Street). Air pollution, particularly sulphur dioxide (SO_2), is known to affect the colonisation of stone surfaces by lichens. The colour of available lichens provides an indication of air pollution; as such, lichens have been used a biomarkers and can convey air quality. Their colouration can be employed to decipher which species of lichens are evident and, thereby, whether they are pollution-resistant species (or not). This is useful information in the absence of air quality stations, such as for assessing air quality on unmonitored roads.

Case Study: Quantifying Lichen Colonisation on a String Course at a Roadside Location in Central Oxford, UK

Abstract: In a conjunctive study of biological colonisation on a string course at the University Church of St Mary the Virgin, close-up digital photographs were taken with a ColorChecker chart on a cloudy summer day around midday. The previously developed outdoor integrated digital photography and image processing (O-IDIP) method was employed in this research in order to quantify lightness and colouration across the top of the string course. Two main types of lichen colours were evident, including whitish (light grey) and yellowish lichens. These colours affected the results, particularly of lightness measured on this south-facing surface. Qualitative assessments of the impact of orientation are considered alongside caveats of the study. Microenvironmental variables remain crucial considerations, especially of sunlight (affected by aspect) as well as local pollution concentrated through an urban canyon.

INTRODUCTION

There is an established literature on lichens used for various scientific purposes. For instance, lichenmetric analysis was employed in the derivation of weathering rates in order to solve basic archaeological problems (Broadbent, 1990). In a study, 350 tombstones with known dates were examined to test the estimated lichen growth curve, supporting the use of the mean of the *n* largest lichen diameters as one of the reliable methods of moraine and tombstone dates (Jomelli *et al.*, 2007). Innes (1985), for instance, selected the five largest lichens on gravestones in Highland Scotland in another study employing lichenometry and, establishing an index, discovered that substrate lithology and orientation (aspect) did not affect ratios of thalli thickness, but that substrate age mattered. There were also spatial patterns revealed in the research that were accounted by onshore winds and the distribution of air pollution, which did not affect results for work performed in Arctic-alpine areas. Evans *et al.* (1999) attempted to establish a growth rate prediction curve for Iceland, also based on the average of the five largest lichen diameters. Lichen diameters (along with documents and radiocarbon dates) allowed for the dating of stones on a Norse fur-trading trail into Sami country (Bergman, 2007). Bradwell (2010) examined lichen growth of *Rhizocarpon geographicum* in north-west Scotland between 2002 and 2009, employing repeat photography (under identical environmental conditions) to study growth rates. He found that larger thalli (10-30 mm) grow faster than those >10m (Bradwell, 2010). A growth curve specific to Corsica, based on information derived from megaliths, created a record spanning 2,000 years, with the curve being more similar to Mediterranean environments than for Arctic and alpine regions (Gob *et al.*, 2003). Lichenometry has been used, for instance, on alpine glaciers, such as on the Murtèl rock glacier in Switzerland, as an additional (dating) method (Burga *et al.*, 2004), and deployed in more recent work by these researchers on the Morteratsch glacier in Switzerland (Burga *et al.*, 2010).

Lichens are known to be sensitive to air pollution. It has been found, for instance, that their lead (Pb) isotopic composition enables quantitative monitoring of atmospheric Pb (Carignan & Gariépy, 1995). Indeed, long-term biomonitoring of nine elements (Cd, Cr, Ni, Pb, V, Cu, Zn, Fe, and Al) with an epiphytic lichen, namely *Xanthoria parietina*, over the course of 7 years in the Adriatic, Italy found it to be a reliable assessment tool for correlation analysis and to decipher trends, as in the composition of air pollution as well as their origin (Brunialti & Frati, 2007). Similarly, in London, 19 species of epiphytic lichen were correlated with traffic-related pollution, namely NO_X, and 16 with bark acidity (Larsen *et al.*, 2007). In the border between Norway and Russia, pollution (particularly of high concentrations of SO_2) increased in the ground air layer between 1973 and 1988 removing vegetation (of lichens and mosses) on rock surfaces, making lichen vegetation barren (Tømmervik *et al.*, 1998).

Researchers have recommended *in situ* biomonitoring incorporating lichens (as well as higher plants) to characterise air pollution in the absence of monitoring stations (Carreras *et al.*, 2009). For instance, at an urban site near Naples, Italy, four environmentally significant elements, namely Fe, Cu, Zn, and Pb, were measured and the most suitable bioaccumulators for these elements were discovered to be *F. caperata* followed by *P. chinense*, with *N. oleander* as a heavy metal biomonitor (Garty *et al.*, 2002). The desert lichen *Ramalina maciformis* was found to be a sensitive species having the potential to be a suitable bioindicator of pollution, with those lichens transported to polluted sites containing more mineral elements after just 7 months (Aprile *et al.*, 2010).

The coincidence of having a string course along with air quality monitoring on the same street provided an opportunity to study the colour-based appearance of lichens for biomonitoring of air quality at a roadside location. The central aim of this investigation is, hence, to execute an outdoor image-based quantification of biological markers to air pollution (biomonitoring) at a location where air pollution is monitored. It may be possible for this environmental monitoring technique to be applied at unmonitored locations as a low cost quantitative approach. Specific study objectives are: 1) to provide data summaries of air pollution at this station, 2) to quantify the level of surface brightness (lightness), and 3) to quantify surface colouration by lichens as proportion values. Finally, caveats are conveyed for the application of the environmental monitoring technique presented in this study.

METHODS

The fieldwork was conducted on 23 June 2011 at the University Church of St Mary the Virgin (2009) located on High Street in central Oxford, UK (Fig. **1a**). The site was selected for its string course, which extends along the entire frontage of the building along a major road (High Street) in the city centre. It was built when the entire nave was rebuilt between 1485 and 1510 (R. Rundle, personal communication). This string course was optimal for its south-facing aspect (orientation) and roadside location (less than 3 m away from this main road) as well as its proximity to an air pollution monitoring station (less than 15 m on the same side of the street eastward from the study area). It is situated near an air quality monitoring station on High Street at All Soul's College (see Fig. **1b**).

The Oxford High Street automatic monitoring site (OS grid reference: SP517063) is situated at a roadside location in front of All Souls College on High Street in Oxford (see Fig. **1b**). Monitoring began at this site since 23 June 2003 of oxides of nitrogen (NO_X) and particulate matter (PM_{10}). Traffic is restricted in this area between 7.30 am and 6.30 pm, with allowance only for buses and taxis during this time. The sampling intake took place 2 m from the road and 1.5 m above the pavement. The Oxford AUN automatic monitoring site is a roadside station located in the town hall at St Aldates (OS grid reference: SP524062) that was active on 15 April 1996 and monitors carbon monoxide (CO), SO_2, and NO_X. Traffic is restricted in the area between 7.30 am and 6.30 pm, allowing access only to buses and taxis. The station is situated c. 2 m from the kerbside and 3.5 m above ground. An urban background site is also located centrally at St Ebbes First School on Whitehouse Road (OS grid reference: SP512054). This site is not considered in this study, as an urban background and is not a roadside location, but has been tracking NO_X, ozone (O_3), and PM_{10} since 09 July 1997. From the Oxford Airwatch website (Oxford City Council, 2014), archive data were obtained from annual data summary sheets for the three monitoring stations between 1997 and 2011.

Figure 1: a) Location of the study site at the University Church of St Mary the Virgin, b) along with one (Oxford High Street) of the three air quality monitoring stations located in central Oxford.

Digital photographs were taken close-up of the entire string course facing south along the top of the structure. The first image was of a portion of the structure that was mostly clear of a lichen cover and was used for comparison of the appearance of lichens eastwards along the remainder of its length. Photographs were taken in cloudy conditions around noon in order to avoid shadows. Images were taken using a Nikon Coolpix S4 digital camera set on a tripod. Camera settings were in Museum mode, without flash and with macro on. A ColorChecker chart

(GretagMacbeth Colorchecker[TM] Image Reproduction Target) was incorporated in each set of photo-pairs for calibration of outdoor lighting and colour conditions. Calibration was performed according to the procedure outlined by Thornbush (2008b, 2013b), with adjustments made to lightness and both colour channels.

More specifically, JPG images (set standard for most cameras) taken at 300 dpi were converted to TIFF and modified from RGB to Lab Color Mode in Adobe Photoshop (Version 7.0). Neutral grey 5 (22) using CIE *Lab* illuminant D50 was used to calibrate the colour photographs with the colour chart (Fig. **2a**). This represented values as follows: $L = 50.867$, $a = -0.153$, and $b = -0.27$, which were simplified for adjustments in Adobe Photoshop as $L = 51$, $a = 0$, and $b = 0$ (Fig. **2b**). After calibration was performed to both brightness or lightness (L) and chromatic (a and b) values, the Histogram function was selected for quantification of the 82 images (Fig. **2c**), spanning the entire roadside frontage facing onto High Street of the string course.

The outdoor integrated digital photography and image processing (O-IDIP) method developed by Thornbush and Viles (2004a, 2004b) and more recently by Thornbush (2008b, 2010a) was, thus, used to quantify image colouration for the assessment of a colour-based (*Lab*) representation of lichen cover. Outputs were originally out of 255 and were converted to proportions. This quantification system conveys increasing lightness (0-100%) and chroma, including green-red (green = 0 and red = 100%) and blue-yellow (blue = 0 and yellow = 100%) spectra. Any indication of lichens could be measured by increasing lightness (due to light-coloured whitish, or light grey, lichens) as well as yellowing of the surface (due to yellowish lichens). Moss could be detected growing on some parts of the string course, including brown mats of desiccated moss, which increased the red and yellow chromas.

RESULTS AND DISCUSSION

Air pollution data in the city centre (at St Aldates) monitored since 1997 (to 2011) appears in Fig. **3a**. NO_X, including NO and NO_2 were all measured locally, with NO being the most recently monitored (since 2005, except for 2006). Monitoring of CO and SO_2 ended in 2007 at this station. At this site, it is evident that pollution was reduced noticeably in the first couple of years. Since 1998/1999, pollution levels have stabilised, with some further reductions in CO, which was relatively high in 2003 as was NO_X. More recent measurements of NO approximate patterns in NO_2. Since 2003, on High Street, pollution levels were generally reduced in the first year of observations, and stabilised since 2005 (Fig. **3b**). Again, here, NO trends follow those evident for NO_2.

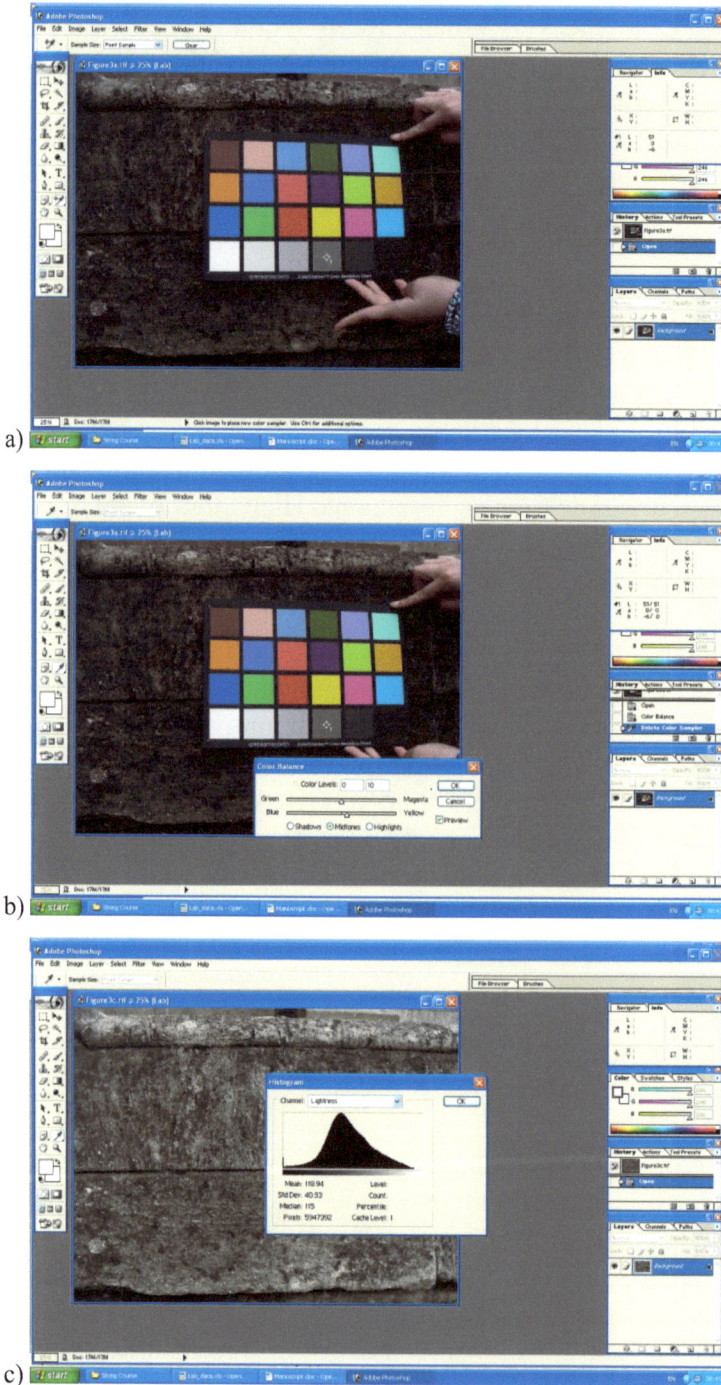

Figure 2: Calibration procedure using neutral grey 5 (22), including a) use of colour chart, b) adjustments, and c) histogram output.

a)

b)

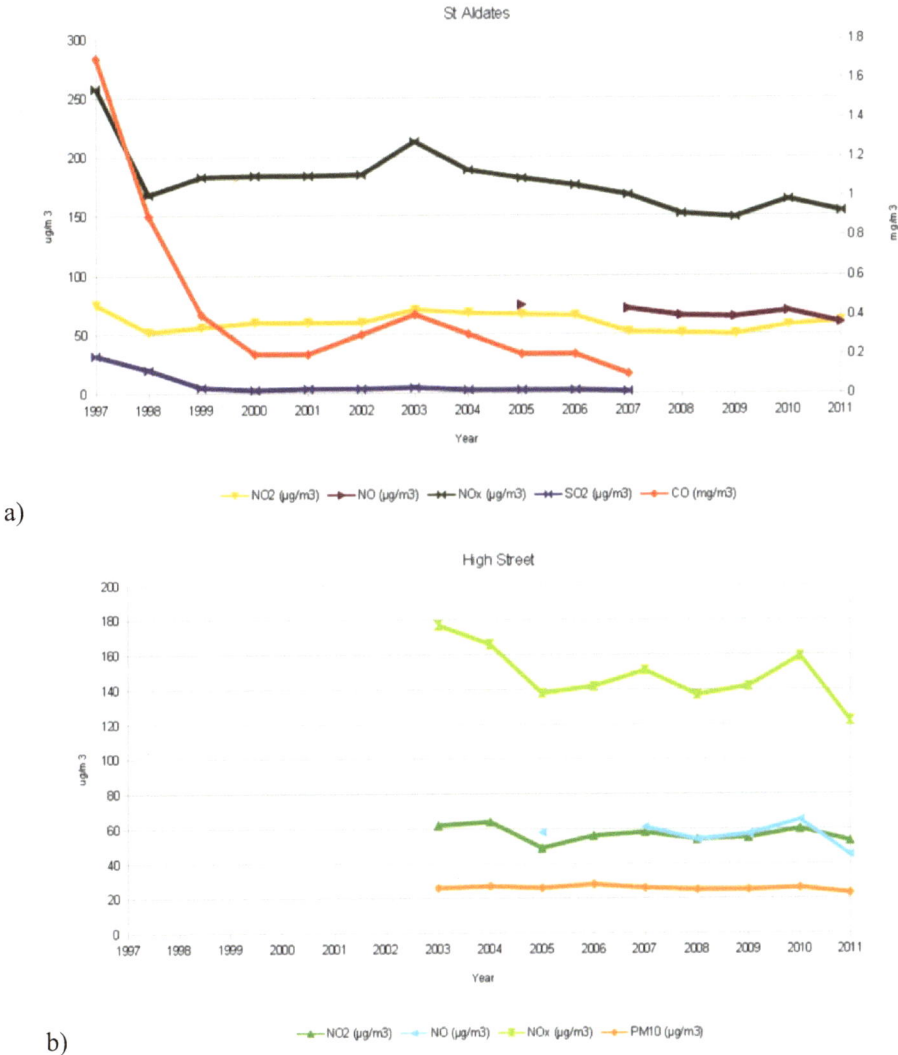

Figure 3: Air quality monitoring a) since 1997 for the Oxford city centre and b) since 2003 for High Street.

The average Mean L varies from 40.82% (at sampling point 53) to 55.67 (at sample 8). This lightness distribution appears to be sinuous from left to right across the string course (Fig. **4a**). The respective Std Dev of L ranges from 12.67% (at sample 46) to 19.05% (at sample 28). It appears to be lowest towards the middle of the string course (from 34 to 50). The Std Dev of a and b are comparatively low (less than 1% for a and 3% for b), with the latter only slightly greater than the former (Figs. **4b-4c**). Mean values of a are close to 50% (highest value of 51.54% at 34) and slightly greater for b (highest value of 53.45% at 2).

a)

b)

c)

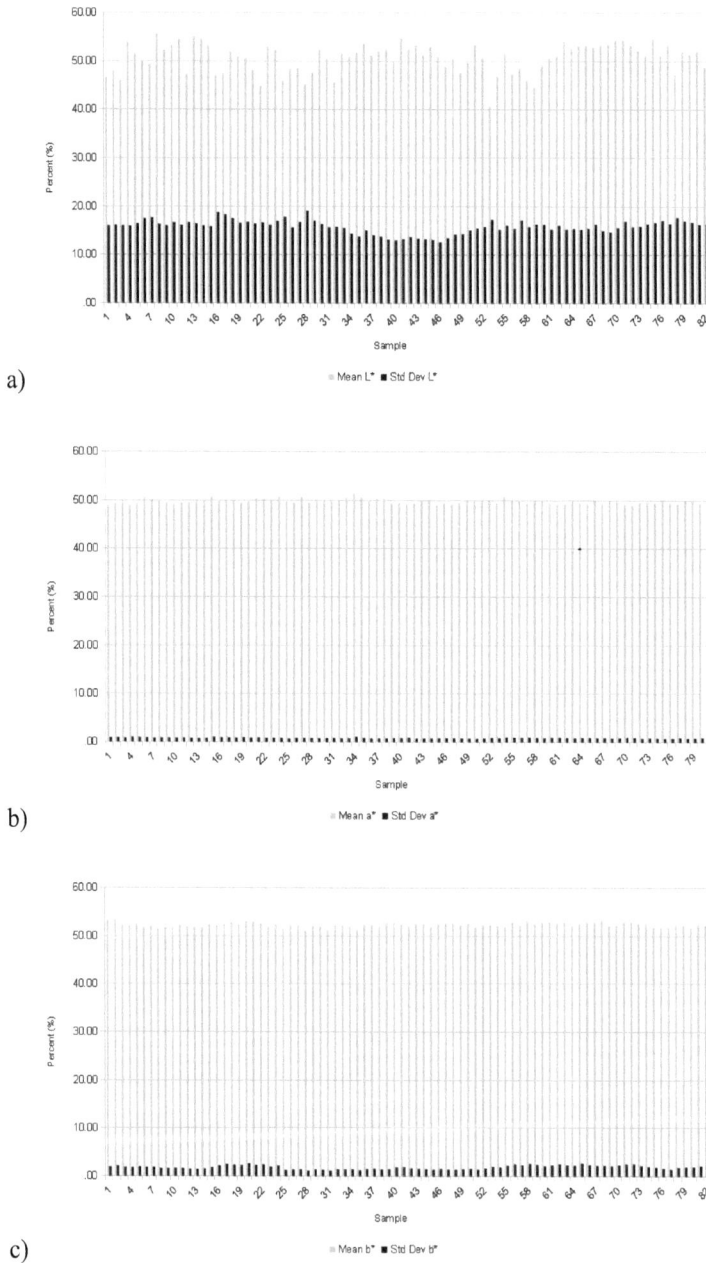

Figure 4: Results for a) *L*, b) *a*, and c) *b* across the length of the string course (left to right).

Results for lightness (*L*) oscillate across the string course around the 50% mark. This visible pattern (in Fig. **4a**) is likely to be produced by the distribution of light-coloured (whitish-light grey) lichens rather than a remnant of local lighting conditions. The reason for this is that Std Dev values, which normally indicate areas of shadow, do not oscillate in the same way. Indeed, towards the middle of the string course, where lightness increases, Std Dev values for L are low; conversely, there are instances where Std Dev values for L increase (peak) where Mean L values are lower (*e.g.*, at 6-7, 16-17, 25, 28, 53, and 78).

Even though there are no visible red lichens, the a spectrum (of green to red) is greater (or redder) at certain sampling points along the string course (including at 6, 15, around 34, and 54). This could be due to brown colouration evident on the structure attributable to moss growth (Fig. **5**). A yellowing of the stone surface can also be seen (in small oscillations) across the string course, with a similarly oscillating Std Dev of b, particularly towards the sides. This colouration is due to the presence of yellowish lichens visible in the photographs (Fig. **6**).

Figure 5: Moss growth on the north-facing side of the string course.

Figure 6: Colouration (and diversity) of lichens seen close-up on the south-facing face of the string course.

The only explanation for the sinuosity (oscillations) in the datasets, particularly for L and b, could be explained by local environmental factors, since all images were calibrated. For instance, there are trees towards the right side of the string course that do not appear towards the left (Fig. **7**). The middle portion of the string course is also exposed (unsheltered by trees), where Std Dev values for L and b are lower, possibly reducing shadows in the former. Nevertheless, it is possible that an open space (free from the sheltering of trees) is creating a more varied distribution of lichens in the middle of the string course. However, the left side is devoid of any tree cover (and is also exposed).

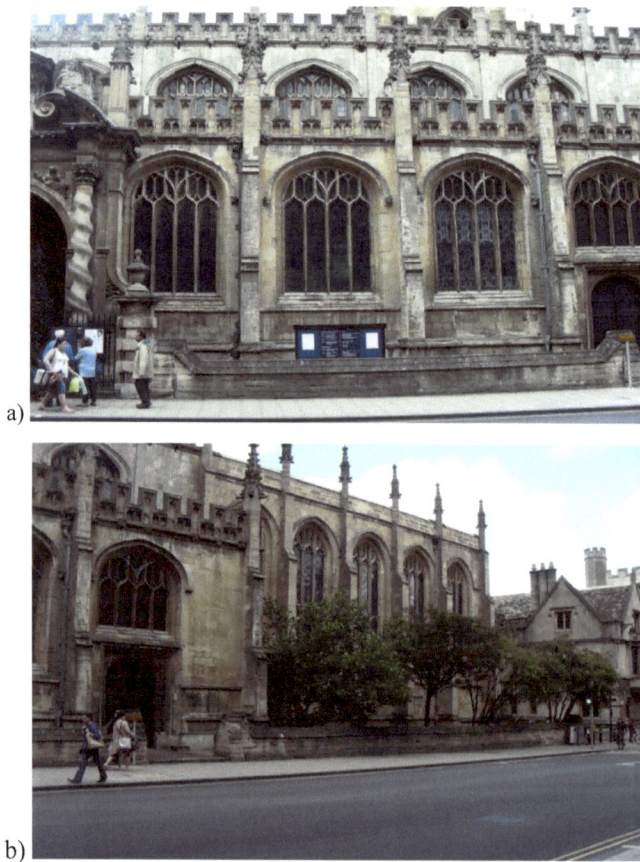

a)

b)

Figure 7: The two sections of the string course towards a) the west (left) and b) east (right).

There do appear to be moisture gradients across this string course. Their north-facing sides, for instance, are greener in colouration due to algal and moss growth (Fig. **8**). In comparison, south-facing surfaces are covered with lichens (mainly light grey, yellow, and some green), but not the other (north-facing) side of the string course. This conveys that the growth of these lichens is not limited by moisture, but rather by direct sunlight (and, indirectly, temperature). The whitish (light grey) lichens seem to be particularly sensitive to lightness levels, growing with a greater proliferation on exposed sections of wall. Aeration should also be considered, especially since grey-green patinas are known to develop in more ventilated areas, such as of the lichen *Dirina massiliensis* f. *sorediata* at the Crypt of the Original Sin in Matera, Italy (Nugari *et al.*, 2009).

Figure 8: Algal and moss growth on the back (north-facing) side of the string course.

Air pollution is known to limit the diversity of lichens to those more hardy species that can survive in a harsh polluted atmosphere. Along High Street (near Magdalen College, for instance), traffic congestion has been declining since the late 1960s (Thornbush & Viles, 2005, see their Fig. **1**, p. 46). This means that there is less traffic pollution from fewer vehicles on this roadway, which is supported by air pollution data (for CO, SO_2, and to some extent NO_X) in the Oxford city centre, as especially denoted at St Aldates. Nevertheless, the number of lichen species currently growing on this string course (*e.g.*, light grey, yellow, green), which could mean that memory (or inheritance) effects could be in-place (cf. Warke, 1996), perhaps associated with surface acidity limiting the proliferation of lichen species. The light-coloured whitish (light grey) lichens in particular appear to be more sensitive to air pollution and could be seen in growing greater proliferation on walls in back (quiet) roads away from busy traffic (Fig. **9**). The darkening of the surface colouration on this string course along High Street makes it difficult to locate lichens; although some are highly visible, such as the light-coloured (light grey) lichens and the yellow types. The latter can be seen growing up to the size shown in Fig. **5**. Lichen growth could be stunted where there is greater acidity on their growth substrate (the string course).

Figure 9: Lichens growing on a wall located in a quiet back street (Queen's Lane).

This study reveals that, although levels of air pollution are a relevant consideration in the quantification of lichens, environmental factors are also important to consider. The appearance of light grey lichens, for instance, seems to be affected by shading. Biomonitoring studies cannot consider air pollution alone without controlling for local environmental factors (microclimate), like sunlight. Monte (1991), for instance, discovered, in the presence of humidity, solar radiation intensity to be a main factor contributing to the decay of the Anio Vetus aqueduct in S. Gregorio near Rome. Aspect is crucial in this regard, as shown in this study, where north-facing surfaces of the string course are clear of lichens and dominated by algal and moss growth, bird droppings, and more. On the other hand, the south-facing surfaces studied here convey light grey and yellow lichens growing on a west-east axis (although not on the north side of the string course). Aspect would, of course, control sunlight (*versus* shading) and moisture, with wetter parts of the wall orientated northward. Lichens growing on cobbles in the Negev Desert, Israel, for instance, over the course of 15 years of observation (between 1991 and 2006) were more plentiful on the north-facing slope (and were aspect-dependent), where they also had a faster regeneration time of c. 45 years

(Kidron & Temina, 2010). In Tasmania, the lichen genus *Siphulella* occurs on stones (as well as soil and peat) in areas of high rainfall (Kantvilas *et al.*, 1992). Moreover, in addition to considering air pollution, urban climate is relevant for lichen growth, especially since lichens are poikilohydric and derive most of their moisture and nutrients from the atmosphere (Jovan & McCune, 2005). This has been conveyed in other research (in the Mediterranean), where northern slopes were dominated by biotic components, including vegetation and biogenic crusts (Kutiel *et al.*, 1998).

Caveats

Lightness variations (consistency of lightness spanning the study) have been controlled by cloud cover (on a cloudy day). Nevertheless, it is possible that some variations in cloud thickness may produce differences in lighting across the span of the string course. It is noteworthy that all images were calibrated and the potential for this error eliminated in the calibration process. It could be assumed, however, that the light colouration of lichens (*e.g.*, light grey) can augment the brightness value where there is a concentration of growths (either in individual size or scatter of growths). Overall, such lichens may increase the measured lightness of the surface. There appear to be smaller deviations in chromatic variables (especially across the green-red spectrum of *a*). Algal coatings could be greening the string courses so that any green lichens will be muted in their representation. Moreover, pulls towards the red portion of the *a* spectrum could produce deviations towards the mean (of green *versus* red growths). Similarly, any discolouration due to iron accumulation on the stone surface could distort the quantification of yellow lichens (and increase *b* values).

It has been observed, since environmental factors may affect air filter sampling (as for particulate matter), that bioindicators like lichens be used in combination (Rossbach *et al.*, 1999). However, it is noteworthy that the rock substrate may also affect the species of lichen growing on rocks, as with glacial erratics found in Greenland, where rock surface properties (nitrogen- and iron-bearing weathering crusts) could be influencing lichen growth (Hansen *et al.*, 2006). It is also possible, in addition to crusts, that the mineralogical composition of rocks could affect the appearance of lichens on this substrate, as observed by these authors for *Orphniospora moriopsis* (which prefers syenitic rocks, with elevated magnetite and phosphorus).

A great diversity of lichens was not evident in this polluted environment. However, had that been the case (had there been a proliferation of lichens and a variety of colours), this would complicate the results along the same colour channel or spectrum (*e.g.*, green-red). For instance, the visibility of older lichens that are desiccated or have been darkened by roadside pollution could also complicate a colour-based assessment. It is noteworthy here that damage can ensue to a lichen's cellular membrane due to accumulation of Cu, Ni, and Pb (especially Cu and Pb, as measured in lichen thalli), but not due to Zn concentrations (Carreras *et al.*, 2005). For epiphytic lichens located in an urban area of Turin, Italy, NO_X and total suspended particles were found to be the main air pollutants affecting lichen diversity (Isocrono *et al.*, 2007). At an urban site in central Argentina, nine species of lichens, with a predominance of *Physcia undulata* and *Physcia endochryscea*, the central area of the city, where there is poor air quality, is a lichen desert (Estrabon *et al.*, 2011). Elsewhere, as in Garhwal (Himalayas), a low lichen diversity was found at polluted sites, where the lichen family Physciaceae dominated; whereas, in a pollution-free area in the city of Pauri, there were 46 species of lichen taxa and a maximum lichen diversity (Shukla & Upreti, 2011). For a nearer comparison, the city of Sheffield, UK was reported to have 77 lichen taxa with an average of about 15 species, with the highest lichen richness appearing on stone substrates (Smith *et al.*, 2010). Finally, this study depended on a string course along a major roadway in the city centre. Qualitative observations and comparisons can be made with other string courses (with a south-facing aspect) located in areas of relatively lower levels of air pollution.

CONCLUSION

It is important to consider growth factors affecting lichen growth situated in the local environment. In this study, this has included air pollution (and acidity of the growth medium or substrate) as well as sunlight (governed by the north-south orientation of this string course) and moisture. Sunlight seems to be particularly important for the appearance of light grey lichens evident on south-facing surfaces (and even in other directions, such as east-facing, but not north-facing). Air pollution is also an important environmental factor for consideration in lichen studies on walls and the size of lichens (stunted growth, as well as the species diversity) and the visibility of older growths may be impacted on by the acidity of the growth substrate.

This study has demonstrated that it is possible to engage a simple technique in order to cheaply capture colouration-based change associated with a lichen cover. These lichens can be used for biomonitoring of local air pollution in the absence of equipment to monitor air quality. The Oxford data convey an improvement in

CO, SO_2, and to some extent NO_X after the implementation of a transport strategy to reduce traffic congestion in the city centre. However, some caution is needed when using lichens for biomonitoring of air pollution due to local environmental factors that may themselves promote lichen growth, such as microclimatic effects produced by aspect as well as local vegetation (trees).

ACKNOWLEDGEMENTS

We are grateful to the Vicar and Ruth Rundle, Parish Administrator, at The University Church of St Mary the Virgin for permission to take the photographs and assistance.

Chapter 5: Wider Applications

CHAPTER 5

Wider Applications

Abstract: It is important to consider wider applications of photography and photographic applications. In this chapter, various developments are considered, including the derivation of an environment history through the photographic record as well as technological applications, as through microscopy and image capture. In this way, the environment can be captured at various scales and from field-based studies to the laboratory.

Keywords: Environmental history, image capture, landscape change, microscopes, scale.

Photography has many applications within geomorphology. Thornbush (2013d) has written on the development of photogeomorphology as a subdiscipline within geomorphology that incorporates the use of photography to study landforms and landscape change. She employed studies conducted in the city centre of Oxford as a focal in her broader literature review since the 1960s. The author postulated that, as evident in the Oxford studies, the subdiscipline should be less fixated on landscape-scale (aerial) approaches and also employ close-up ground-based photography and rephotography in the assessment of landforms and landscape change. This broader scale of application could benefit the study of stone soiling and decay (weathering) studies as smaller forms may be overlooked.

Recent work by Thornbush (2014) used a historic record of 1,123 archival photographs from English Heritage, namely ViewFinder, in order to consider the effect of urban greening on urban climate. Deriving 69 years of record (from 1860 to 1929) of the cover of climbing plants on buildings and structures located in central Oxford, she discovered effects of such a vegetation cover in the built environment. Specifically, in 1880, mean temperature was lowest (9.16°C) at a time when there was the most evidence of climbing plants (195 photos). Conversely, mean temperature was greatest (9.6°C) in 1920, when only six photographs depicted climbing vegetation. There was also a temporal trend evident for potential evapotranspiration, which was lowest in 1880 (617.88 mm) and greatest in 1920 (629.19 mm). The latter was controlled chiefly by mean temperature (affecting evaporation) rather than a vegetation cover (affecting transpiration). Similar trends were not found for total precipitation. Whereas both mean temperature and potential evaporation showed a negative linear covariance with the appearance of climbers in ViewFinder images ($r = 0.8993$ and 0.9679,

Mary J. Thornbush and Sylvia E. Thornbush

respectively; both statistically significant at the 0.01 level, two-tailed distribution), the total hours of bright sunshine parameter conveyed a positive linear relationship ($r = 0.7179$; statistically significant at the 0.05 level, two-tailed distribution). This latter finding suggested, because climbing vegetation seemed to actually reduce potential evapotranspiration through a cooling effect (of reduced mean temperature on average per decade), that urban greening promotes fewer clouds (and more bright sunshine under a clear sky condition). This cooling effect of vegetation (through shading, but also through the appearance of plants) operates to reduce the urban heat island effect, which can contributing increasingly to global warming through increased urbanisation on a global scale.

Indeed, photographs can record images that portray weathering forms at various scales, especially now that many instruments have visualisation and image recording potential. Scanning electron microscopy (SEM) has photographic capabilities, whereby images can be captured of microscale features on stone samples. Thornbush and Viles (2006), for example, used SEM to examine Bath limestone samples exposed before *versus* after the OTS. They discovered that within hollows there was much microbial growth on stone surfaces before the OTS, with evidence of spreading growth away from hollows after the OTS. There was more pronounced soiling evident at Longwall Street, where microbial cover was lower. The reverse occurred at High Street, where there had been traffic abatement and surface lightness was much improved.

Thornbush and Viles (2007b) also used SEM micropictograms to depict dissolutional features on weathered *versus* unweathered surfaces of Bath (Box Hill) limestone. Their experiment in a climatic chamber revealed that stone samples placed in a strong carbonic acid (H_2CO_3) solution experienced a greater amount of dissolution. This was conveyed in water samples by a higher concentration of total hardness and Ca^{2+} and the appearance *via* SEM of more microscopic dissolution features. These findings suggest that the enhancement of atmospheric CO_2 under global warming, provided that there is an adequate supply of moisture, is likely to accelerate the rate of dissolution, especially of newly replaced calcareous building stones. However, surfaces that have been previously exposed and weathered, such as those exposed for a century or more, may be less susceptible to enhanced rainfall acidity. In consequence, conservation techniques that remove weathered surfaces, including stone cleaning, may actually accelerate the decay of historical limestone structures in the future as concentrations of carbonic acid are enhanced.

Using a digital camera to capture free-style photographs close-up with a pointer for scale, Thornbush (2012) was able to develop a weathering scale for limestone. Based on a photographic survey, she was able to capture various images conveying the different weathering forms evident in the Oxford city centre. These were mostly intact along Queen's Lane, where buildings and structures have not been cleaned or restored and represent the normal progress of soiling and decay in the history of Oxford. It was possible to discern discrete features at the mm to cm (micro- to meso-) scale, including pits, flakes, scales, spalls, blisters, and caverns, which were mostly associated with chemical weathering. The scale also enabled capturing biological weathering (bioweathering) from flora and fauna as well as levels of soiling.

In this first part of this eBook, consideration has been given to photographs to pictorially or quantitatively depict landscape change (both temporally and spatially), as part of an urban geomorphology. In the next part, focus will shift to cultural change within urban landscapes. Here, photographs are employed to develop records of cross-temporal change, where the archaeological record is involved.

ACKNOWLEDGEMENTS

Declared None.

Part II: Capturing an Urban Landscape

S. E. Thornbush

School of History, Classics and Archaeology, University of Edinburgh, William Robertson Wing, Old Medical School, Teviot Place, Edinburgh, EH8 9AG, Scotland, UK

> A photo may be seen as a fact, as evidence that an event occurred or that a building looked a certain way at a given time. Yet the powerful document of the photograph is a product of a split-second. Each image, directed by the individual behind the camera, reflects the photographer's motives for wanting to record that scene (Chabot, 1996, p. 574).

Monuments and other similar structures of either religious or cultural representation have been of interest in the UK for many centuries. Not only do these structures represent truths and common interests, but they also establish a sense of self-individuality or a sense of being within a community. In as early as 1559, the Queen Regent issued a proclamation for the preservation of sacred structures (BHO, 2014). This alone demonstrates the worth of such monuments then, but it should also pave the way to understanding the value of monuments in contemporary society. The fact that, in the UK, there are several urban churchyards with some *in situ* headstones suggests that these structures represent more than mere history, but a heritage of a people that once lived in these cities.

Headstones have also been captured on print in literature. In Gothic novels, the churchyard has become a necessity alongside the burning candle in the night in exploration of the castle grounds. Wordsworth, in the mid- to late-19th century, was asked to write about the cemetery to encourage the living to see these deathscapes as peaceful and inviting landscapes, an oxymoronic view of what churchyards had become with overflowing bodies with little to no decay of the dead (Sanchez-Eppler, 1988). Henceforth, Charles Dickens's view of English life must have been targeted by the morbidity associated with the crowded churchyards in the late century. However, authors such as Susannah Moodie did not hold the same views of death; she saw it as a change, almost like an expectation brought on by time (McKendry, 2003, p. 3).

The outcome of death, and its celebration, is evident in the urbanscape containing a collage of architectural structures, edifices in their own defence, that act as emblematic structures found in churchyards, burial grounds, and cemeteries. Each monument has a unique representation of the commemorated and this uniqueness is what drives the need to explore these locations. These landscapes are useful for many reasons, the main one here is for remembrance. However, other reasons include: education, religion, manufacturing practices, masonry, stone types, architecture, art history, and geography, to name but the most common reasons to study headstones.

Through photography these types of structures could be prolonged; the existence of such socially, educationally, and personally important media could continue to be appreciated. Having images of headstones available can sustain research in particular, and, thus, the understanding of a part of our cultural heritage. Capturing a scene through the lens has proved to be a very valuable tool, unlike the paintbrush that would require time to produce an image of any desired scene. Photographs capture important moments and sustain important elements through time, in particular a digital photograph. Research that uses photography has great potential.

Photography also promotes more in-depth research rather than a rapid investigation of an object. For headstones research the importance of relying on the photograph to aid in the research process is that there are several elements at play whilst conducting research in the churchyard. Time is a main factor; if a churchyard has anything over 100 headstones and several measurements need to be considered alongside the recording of the inscriptions and the condition of each headstone, the researcher would not be able to go through more than half of the existing count. With dimensions, Cardinal readings, inscriptions, epitaphs, weathering features, type of stone, and other elements, such as vandalism and damage to the stone, for one headstone alone 10 minutes to a quarter of an hour would be the required time spent on recording, depending on the size of the headstone. Having access to photography in the field means that, although measures, dimensions and cardinal readings, must be taken as usual, and the location of the headstone is plotted on the map of the site, the photograph of the headstone could be revisited and additional details could be recorded outside the field, halving the length of time needed in the field.

The following chapters will delve into the different levels of the importance of photography. Case studies have also been incorporated to demonstrate how photographs can be used to facilitate research. These case studies are mostly all based on the second-listed author's own research and experiences employing photography in her headstones research.

Chapter 6: Establishing a Visual Record

Establishing a Visual Record

Abstract: The use of photography in research has enabled the preservation of several important features historical monuments such as headstones. This established visual record creates a gateway to studying the past in a more accessible way. Yet, these have also been used for more sentimental reasons, in the remembrance of the dead; by establishing a seriation in their use on headstones, which commence in the mid-1980s and then in the 1990s in Oxford, UK cemeteries.

Keywords: Cemeteries, headstones, photograph, portraiture, seriation.

Capturing headstones digitally (or otherwise), particularly historical ones, has now become more of a need than ever before, especially with the ongoing decay surrounding any monument that experiences constant outdoor exposure to the elements in a polluted (urban) environment. These breakdown stones degrade the epithets that family members have inscribed in their memory. By capturing the monument digitally maintains the survival of the image. In some cases, material culture like headstones are captured on film and, in this way, are stored for as long as that form of media survives. For example, with the film *The Prime of Miss Jean Brodie* (1969), in one scene Miss Brodie is walking in Greyfriar's kirkyard in Edinburgh and in the background the headstones that were erected in the late 1960s will always be preserved as long as that film remains accessible.

However, not only are the elements at play here, but the churchyard is a changing landscape. It is not as stable or quiescent as one would think. The simple fact that it is a place of burial and commemoration does not entail that the scattered monuments in the typical churchyard will remain as such. In a mere year's time, it is possible to see the minimisation of the number of headstones in a given churchyard. In fact, in the authors' research, it was impossible to locate at least one headstone after a return visit only a year after a photographic survey was conducted at Greyfriar's kirkyard, Edinburgh. There could be several reasons for this, one of which could be due to management practices, whereby issues, such as safety, come into play. If a headstone is unstable, it is tagged automatically as a hazard and must be either flattened or removed. In many cases, unstable headstones are flattened onto the ground to avoid injury to anyone. In other cases, the headstone is relocated to an area where less damage could be done so that it can be secured either against the church or the churchyard wall.

Unfortunately, what this means is that this information is lost, or at least would have been lost had it not been previously recorded. In terms of church records, these can be very limiting to research as they only provide very basic details of the interred. These details do not include any architectural features, whether the headstone had a motif or not, what introduction was used, or even whether an epitaph was inscribed that best described the deceased, let alone the different font types inscribed on the face of the headstone. This is where a visual record becomes an asset, particularly in historical archaeology.

In the same way, in order to preserve a memory, photography became the primary source of doing this. The portrait became of interest in several parts of the world from 1839 (Langford, 1978). However, before that time, portraiture had already been established. There are several portrait miniatures displayed in the Victoria and Albert Museum, London, which were popularised in the courts of France and England in the early 16th century (V&A, 2014b, para. 3). During the Victorian period and well after that to the 21st century, photographs or portrait miniatures, were, and continue to be used, on headstones to commemorate the dead. However, this medium has never truly taken flight, and as soon as it starts being used, it disappears from interest so much so that several headstones contain nothing but inscriptions, which is easily found on historical headstones. However, 20th century headstones decreased in size through time, unlike their Victorian forerunners, and, because of this, less space on the surface of the headstone meant that only the recording of important information was made possible. Hence, the use of a photograph to help define the deceased only made sense.

The case study that follows examines the use of portraits on headstones at four different cemeteries in Oxford, England. This study is appropriate here as it conveys the importance of photography in different facets: the use of photography for remembrance and the use of this in research.

Case Study: The Use of Photographs on Headstones at Four Oxford Cemeteries

Abstract: Using photographs on headstones in Oxford cemeteries from the early 1900s was not a popular fad. This form of commemoration entered in and out of the going trend without contributing to its continuity in Botley, Headington, Rose Hill, and Wolvercote cemeteries in Oxford. The results show that more than half, and in some cases more than double, the headstones that used a photograph to honour the dead were commemorating men. However, the average use of photographs on headstones in each of the cemeteries only reached 8% that of the total number of headstones at each site, and in 1985 there were a total of 5 headstones with photographs; this trend continued until 2010 whereby five or more headstones were evident in total.

INTRODUCTION

Prior to the introduction of cemeteries in Oxford, churchyards were the only burial places to think of, and by 1843 every churchyard was full (Jenkins, n.d.). By this time in France, cemeteries had already been established. In fact, by 1789 ecclesiastical control was removed from cemeteries (Kselman, 2014, p. 166). The fact that the English were not in the foreground in burial practices is not surprising as cemeteries were consecrated long after the saturation of bodies in churchyards, which lead to other problems of sanitation and disease. To avoid this from further affecting the masses, Headington cemetery had become a public burial ground in 1885 and, subsequently, land was purchased just outside the city of Oxford in 1889 and 1890 to house three cemeteries: Rose Hill, Wolvercote, and Botley (Jenkins, n.d.). Their sizes varied, with Wolvercote consuming the most acres, followed by Rose Hill, and then Botley. New parish cemeteries were consecrated in 1848 to help house all of the human remains that were overflowing in the churchyards at the time (Jenkins, n.d.). The location of these cemeteries varies, with Botley cemetery being the closest to the city centre at 1.6 miles. Subsequently, Rose Hill cemetery lies 2.4 miles away from here; Wolvercote and Headington cemeteries are both 2.8 miles away from the city centre. In terms of their orientation, Botley stands in the west, Wolvercote in the north, Headington in the east, and Rose Hill in the southeast.

Photography has been used in research for evidence, particularly in historical research (Achterberg, 2007); however, these could be used in any area as long as it is achieves its purpose. Photographic research began in the field of anthropological and social science, but has extended to other disciplines (Fletcher, n.d., p. 2, as in education Wang, 2006). These photographs are used to show, describe and analyse material culture (Jacoby Petersen and Ostergaard, 2003, p. 3), although the use of photographs in research remains sparse (Ray, 2012, p. 288), and is still in its infancy (Tinkler, 2013, p. 116).

For research purposes, photographs are essential as they contain a substantial amount of information; landscapes, material culture, or people could be photographed to be studied away from the field or on site. According to Rose (2011), this type of photography is called 'photo-documentation', which makes the assumption that these photographs are accurate records of what was being photographed (p. 301). According to Jacoby Petersen and Ostergaard (2003), the photograph is very useful as it portrays what is evident, as in using photographs to remember the dead, but also to show 'what is not' (p. 8). The latter element could be useful in correcting a myth or assumption relating to something that could be visually captured. Yet, according to Crang and Cook (2007), there are limits to what could be photographed in social science (p. 107). Moreover, by using photography in research, it is possible for a subjective selection of materials to skew the entire truth about any material culture (Jacoby Petersen and Ostergaard, 2003, p. 8). In this way, the use of photography can be used in either art or in scientific evidence (Schwartz, 1989, p. 120), and this has to be clear right from the beginning to apply this visual interpretation accurately.

In this research, studying the use of photographs of the dead on headstones is a sensitive issue, particularly where publishing is involved. Despite the fact that headstones are meant for public use, there is some sort of respect that still needs to be honoured. In any form of research, the problems associated with publishing personal details found on a headstone could be avoided by isolating for specific features, such as font style, motif use, and, possibly, the use of a photography on the headstone. The latter could be made possible with the use of software to blur out personal details that would otherwise cause problems.

Using photographs on headstones was once fashionable as it was a visual image that described what was most important and what was able to evoke (Linz, 2014) emotions, particularly as photos evoke deeper feelings than do words (Harper, 2002, p. 13), which would be appropriate for the living so that memories are sustained through time. Having a photograph available for the living of the dead, allows the former to reflect on their experiences with the person (or people) with which they once lived. This is also reflected in research conducted by Liamputtong (2007), whereby photographs were used to depict experiences of homelessness amongst the participants; feelings were evoked through imagery that would be linked to such as traumatic experience as this. Frith and Harcourt

(2007) used photography to capture women's experience with chemotheraphy. In this way, the traumatic experience of losing a loved one, would bring back memories of not only the loss but the memories that might have otherwise been lost through time. Moreover, the idea of having an image to reflect on contains much more of a message or a form of expression than if there were just the inscriptions on the headstone to commemorate the dead. Pink (2013) stated that there are things that photographs can express that words cannot (p. 177), and in this research this is the case for choosing a photograph over standard inscriptions. Pink (2012) argued that images should be read just like a text (p. 10); this could be interpreted through lines of the face and eyes and the clothing worn by the deceased. All these together exhibit emotions that are necessary in the mourning process, which, ultimately, is what the purpose of these were in the 20th and 21st centuries.

METHODS

Fieldwork was conducted at four Oxford cemeteries in 2014: Botley, Headington, Rose Hill, and Wolvercote. Headstones with photographs of the dead being commemorated were photographed using a digital camera. The conditions were mostly sunny so that the natural illumination could cast shadows, particularly on headstones with high relief, to create contrast that would then permit the deciphering of the inscriptions through creating contrast. The date and name of the deceased were noted in the field. Post-fieldwork involved inputting the details written up on site onto an Excel spreadsheet and graphs and a seriation created. During this time, Oxford City Council was contacted and provided official numbers of interments (excluding non-headstone burials), and these were graphed against the field results for analysis. It was also contacted regarding costs related to burials at all four cemeteries.

RESULTS

Based on the headstones recorded in this study, it was in Wolvercote cemetery that the first headstone with a photograph was observed (in 1925), but it was difficult to decipher whether it belonged to a male or female. Moreover, in Headington, there was a dateless headstone commemorating a male, although the second commemoration date was 2003, the first one was impossible to decipher. In 1947, in Botley, the earliest headstone commemorating a female was evident, while in 1964 in Wolvercote was the first headstone with a picture commemorating a male. Botley cemetery was the first to have a headstone with a photograph (1947, female). The latest headstone to have a picture commemorating a female was found in Botley in 2012, and this was also true for that commemorating a male.

Table **1** demonstrates that there were a few headstones where the date was unknown or the gender was unclear. Outside of these, for Wolvercote cemetery 28% of headstones with pictures commemorated women, while in Rose Hill cemetery this number was 29%, Headington was 30%, and Botley 48%. The greatest disparity between commemoration between women and men through the use of photography was found in Wolvercote. Of the headstones with unknown dates or sex, there were more in Wolvercote (12%) than in the other cemeteries (Botley, 9%; Headington and Rose Hill both at 4%).

Table 1: Female to male ratio in the use of photographs on headstones

	Female	*Male*	*Unknown*	*Total*
Botley	15	16	3	*34*
Headington	24	57	3	*84*
Rose Hill	20	49	3	*72*
Wolvercote	30	77	14	*121*
Total	*89*	*199*	*23*	*311*

NB: one headstone had no date.

Table **2** illustrates the number of headstones with photographs within each cemetery; the proportion of headstones with photographs was at 7%, and the cemetery that contained the most of these was Rose Hill (13%). The remaining cemeteries did not reach 10% (Headington, 8%; Wolvercote, 6%; and Botley, 4%).

Table 2: Ratio of total burials to those with photographs

	Photographs	Assumed No. Photographs	*Total Headstones*
Botley	34	917	*951*
Headington	84	926	*1,010*
Rose Hill	72	503	*575*
Wolvercote	121	1,913	*2,034*
Total	*311*	*4,259*	*4,570*

Tables **3a** and **3b** show seriations from the results gathered in this research. The former illustrates that in 1985 the fashion of using photographs on headstones began with a total of 5 headstones. The use of photographs here fluctuated throughout the period peaking several times in 1994, and then in 2003 and 2009, while a peak in the number of burials in total was reached in 2002 and 2003, yet these official numbers also fluctuated throughout the period.

Table 3a: Seriation of headstones with photographs

	Botley Nos	Headington Nos	Rose Hill Nos	Wolvercote Nos	*Total Nos*
1925	0	0	0	1	*1*
1926	0	0	0	0	*0*
1927	0	0	0	0	*0*
1928	0	0	0	0	*0*
1929	0	0	0	0	*0*
1930	0	0	0	0	*0*
1931	0	0	0	0	*0*
1932	0	0	0	0	*0*
1933	0	0	0	0	*0*
1934	0	0	0	0	*0*
1935	0	0	0	0	*0*
1936	0	0	0	0	*0*
1937	0	0	0	0	*0*
1938	0	0	0	0	*0*
1939	0	0	0	0	*0*
1940	0	0	0	0	*0*
1941	0	0	0	0	*0*
1942	0	0	0	0	*0*
1943	0	0	0	0	*0*
1944	0	0	0	0	*0*
1945	0	0	0	0	*0*
1946	0	0	0	0	*0*

Table 3a: contd….

1947	0	0	0	0	*1*
1948	0	0	0	0	*0*
1949	0	0	0	0	*0*
1950	0	0	0	0	*0*
1951	0	0	0	0	*0*
1952	0	0	0	1	*1*
1953	0	0	0	1	*1*
1954	0	0	0	0	*0*
1955	0	0	0	0	*0*
1956	0	0	0	0	*0*
1957	0	0	0	0	*0*
1958	0	0	0	0	*0*
1959	0	0	0	0	*0*
1960	0	0	0	0	*0*
1961	0	0	0	0	*0*
1962	0	0	0	0	*0*
1963	0	0	0	0	*0*
1964	0	0	0	1	*1*
1965	0	0	0	1	*1*
1966	0	0	1	1	*2*
1967	0	0	0	0	*0*
1968	0	0	1	0	*1*
1969	0	1	1	1	*3*
1970	0	0	1	1	*2*
1971	0	0	0	0	*0*
1972	0	0	1	0	*1*
1973	0	0	0	1	*1*

Table 3a: contd….

1974	0	0	3	3	6
1975	0	0	2	0	2
1976	0	0	0	0	0
1977	0	0	2	1	3
1978	0	0	0	0	0
1979	0	0	0	0	0
1980	0	2	2	1	5
1981	0	0	2	2	4
1982	0	2	2	0	4
1983	0	0	1	0	1
1984	0	0	1	1	2
1985	1	0	3	1	5
1986	0	1	3	0	4
1987	0	0	5	0	5
1988	0	1	2	2	5
1989	0	1	3	2	6
1990	0	5	5	1	11
1991	0	5	8	1	14
1992	0	2	2	1	5
1993	0	1	3	2	6
1994	0	2	12	1	15
1995	1	4	3	3	11
1996	0	5	0	1	6
1997	0	3	0	4	7
1998	1	7	0	5	13
1999	1	3	0	2	6
2000	0	6	0	2	8
2001	1	5	0	4	10

Table 3a: contd….

2002	2	10	0	3	*15*
2003	4	11	0	2	*17*
2004	4	1	0	10	*15*
2005	2	0	0	7	*9*
2006	0	1	0	10	*11*
2007	0	2	0	9	*11*
2008	5	2	0	7	*14*
2009	5	0	1	11	*17*
2010	3	0	2	8	*13*
2011	1	0	0	3	*4*
2012	2	0	0	2	*4*
Total	*33*	*83*	*72*	*121*	*310*

NB: one headstone had no date.

Table 3b: Seriation of the number of official headstones for each cemetery. Numbers were provided by Oxford City Council

	Official Nos for Botley	Official Nos for Headington	Official Nos for Rose Hill	Official Nos for Wolvercote	*Official Total Nos*
1998	50	137	46	127	*360*
1999	52	140	54	124	*370*
2000	41	117	53	98	*309*
2001	54	69	51	107	*281*
2002	55	121	34	143	*451*
2003	65	81	47	123	*451*
2004	76	55	42	169	*423*
2005	54	48	41	159	*393*
2006	60	44	43	139	*286*
2007	74	28	27	146	*275*
2008	89	43	29	143	*304*
2009	86	39	29	141	*295*
2010	58	30	31	128	*247*
2011	60	26	22	132	*240*
2012	77	32	26	155	*290*
Total	*951*	*1,010*	*575*	*2,034*	*4,975*

Table **4** illustrates when the first headstone with a photograph was noted, and, based on this, the progression follows with Wolvercote, then Botley, and Rose Hill and Headington following close behind.

Table 4: First headstone with a photograph

Cemetery	Year on Headstone (Latest Year)
Botley	1947 (2012)
Headington	1969 (2008)
Rose Hill	1966 (2010)
Wolvercote	1925 (2012)

DISCUSSION

The results were as expected, with a far greater number of headstones without photographs. These types of headstones trended differently in different cemeteries. Generally, the two largest cemeteries (Wolvercote and Headington) contained the most number of headstones with photographs. This is, thus, expected; having the largest cemetery contain the most diversity makes the most sense. In this way, there are more people to make choices about what they want on their headstones, from motifs to the colour of the stone.

In terms of gender, it was discovered that there were more headstones with photographs commemorating men as the first to be interred (69%, based on Table **1**). There could be two explanations for this. First, assuming these men's wives or partners chose or influenced their choice in having a photograph, this could suggest that women were more interested in having their husband's photograph displayed on their headstone for their own purpose or for display to all those visiting their loved ones. This sort of commemoration implies more of a global or shared commemoration, so that they are not the only ones mourning the loss of their loved ones as this would keep their memory alive. To satisfy this argument, Tarlow (1999) emphasised the emotional role of headstones. She stated that their purpose is to be sentimental. Moreover, it is more sentimental to see someone's image on a headstone than solely inscriptions.

This idea is not meant to assume that women are more sentimental than men; in fact, the second argument carries that these men chose their headstones before their death, in preparation for retirement and the like, which is customary. There are some people who pay for their burial plot years before their death, as to make this decision before death is a great responsibility and something they would not want to burden their families with. Following this would imply that men, known for being more visual, would want their wives and partners to remember them as they once were (Hamann, 2004).

Generally, it is very difficult to analyse choice if it is not clear who is making the decision particularly as cemeteries have choice upon choice of motifs and lettering fonts and so on. These available choices would not require much thought. Moreover, according to Stewart (2007), the idea of having a personal headstone was spurred on by romantic ideals and graves provide a place for the living to pay visits to the dead, a relationship that would have a long life. In this way, it would make sense that these men would have wanted their wives' mourning period to be easier by having a familiar face, or perhaps a face to speak with during their visitations.

These choices available in some sort of catalogue might have also had unchangeable options; this means that the cemetery office would have a pre-prepared package. This would mean that choosing a headstone would no longer require much work. The headstones with photographs, all seemed to have a similar design, which could be a coincidence that all those people who chose to have a photograph on their headstones would pre-order the same font, design, and type of headstone.

Oxford in 1985 marked the beginning of the use of photographs on graves and a decade later this style peaked, yet it was not as popular as those headstones that were devoid of photographs. Nevertheless, after this, individual cemeteries caught on with this trend, starting with Rose Hill, then Headington, followed by Wolvercote. The numbers in these were significantly higher, which states that this trended in the southeast then moved up to the east and landed in the north. It circulated through Oxford. According to Smith *et al.* (2011), working class groups have created a commemorative heritage separate from any official memory. These types of headstones were clearly different from the norm based on the numbers in Table **2**, which suggests that this trend was taken up mostly by working-class groups in Oxford cemeteries.

CONCLUSION

The need to remember a loved one through a photograph implies that there is a strong sense of community. People who are very close and are used to seeing their family members at meal times and other common times throughout the day would also need to somehow commemorate their loved ones with a photograph displaying the way in which they once remembered them. This does show strong sentimentality despite the fact that the option of having a photograph on a headstone could have been the easier option through a packaged deal. The Victorians too used photographs on their headstones to celebrate the fact that they could afford to and the fact that they were capable of seeing their loved one in the way they remembered. This concept will never be lost in any society. Memories can be lost if they are not nourished, and photography is a very good way of establishing this. The use of a miniature was also quite similar, as these would be given to loved ones when officers were going off to war. In this way, it could be regarded as a form of commemoration and remembrance, a sentimentality that will never be relinquished.

ACKNOWLEDGEMENTS

Much gratitude is given to Trevor Jackson at the Oxford City Council for providing useful burial information for Oxford cemeteries.

Chapter 7: Visualising Preservation Issues

CHAPTER 7

Visualising Preservation Issues

Abstract: Changes in the way that photographs were taken have led to different methods of preservation. The advent of the digital camera has made photography more accessible as well as easier to use. It has benefited the research method in the field. The preservation of the photograph has also been improved, and methods of storing photographs have also improved. These have all aided in recording key historical information particularly found on headstones. Changing traditions in the use of motifs and epitaphs on headstones in England have been noted through the use of digital photographs.

Keywords: Churchyard, epitaphs, headstone, in-field issues, motifs, preservation, vandalism.

The photograph itself has evolved throughout the years. From its introduction in the late 1830s, its use and storage has changed drastically (Langford, 1978). At first, the idea of photographing a live or inanimate object was a novelty and a very useful tool. The subjects would be prepared to be captured on film and this could have meant hours spent on beautification. The artefact was stored carefully in order to preserve a memory of which could have been a wedding photograph or one taken of an army officer for his wife before setting off to fight for his country. During these times, the preservation of a photograph was very real; however, today, preserving a memory does not take on the same meaning.

According to Nassar (2006), a large proportion of photographers from Europe were working in the Middle East from 1839 to 1900 capturing photographs of Palestine to sell overseas (p. 142). These photographs captured the Holy Land for the pilgrims visiting in the way the photographer wanted. By contrast, the object (or individual) being photographed remained the important element in the captured image, so much so that it could be incorporated into any important issue being addressed well into the future (Nassar, 2006, p. 148). Therefore, incorporating photography into research contemporarily could be highly beneficial to both research and the researcher.

This idea of preservation has taken on a strong link to headstones and churchyard research, yet it could be related to all research that deals with material culture, to include historical buildings and people. The need for preservation stems from, in headstone research, the destruction of headstones through natural forces such as weathering, but also through vandalism and policy. According to Brears (1981), headstones have been used as paving stones or have been removed when churchyards and cemeteries have been cleared for access of space to accommodate fresh burials (p. 92). In fact, in various places in England and Scotland, such as York and Edinburgh, headstones had been ordered removed so that more bodies could have been buried in central churchyards. Therefore, this issue of the preservation of material culture has been a long standing one, and one of the best ways of preserving such monuments is through the use of photographs.

One growing issue that clashes with the idea of preserving monuments is the need for development. With the advent of population growth in cities, more land to build on for housing, in particular, has become a pressing issue. In this way, the result has been 'the clearance of many Historic-period cemeteries' (Stewart, 2007, p. 116). In some cases, the headstones themselves were used to pave the churchyard to be used as a car park (Stewart, 2007, p. 117), as has been noted at St Mary's churchyard in Scarborough by the author. The loss of information about the dead in this case is definite.

DIGITAL PHOTOGRAPHY

Kodak's first digital camera in 1975 was the advent of a new possibility in preserving everything from memories to artefacts, and much later, by 2005, digital cameras were widely popular as was the smartphone camera that gave everyone the opportunity to record anything they wanted, cost-effectively (de Castella, 2012). In this way, any image could be stored safely electronically and easily altered too. This also makes up for lost time, the amount of time it would have taken to capture the image the way intended. There is no longer any need to beautify the subject when through photographic programs, this could be done digitally, from changing the subject's hair colour, to adding a scene to the background. It is easier to create a memory that one dreams of rather than experiences. In this way, the preservation of the photograph is something that has, for the time being, been achieved. However, digital photography is also highly useful to the researcher. As the digital image is now of high quality, it means that a detailed image is captured, minimising the hours needed to spend in the field.

PRESERVATION ISSUES OF HISTORICAL MONUMENTS

In order to continue to capture headstones in photography, it is essential that material culture is retained *in situ*. However, there are several issues that stand in the way of this to include: 1) vandalism and theft, 2) weathering and stone types, 3) in-field issues, and 4) selective preservation.

Vandalism and Theft

There have been several council projects in the UK whereby sustaining the safety of headstones has lead to the damage of such stones. Conducting what is known as the 'topple test' has been applied to headstones in places like Birmingham in November 2010. In Edinburgh, the kirkyards were classified as 'at-risk' in October 2009. According to Mytum (2004a), some monuments were dismantled because they were considered by the authorities to be unsafe (p. 115), and, if this continues to take place, there will soon be no evidence of burial styles from the past. In the author's case, after a year of returning to a specific churchyard, some headstones were missing, having been removed from the location they had once been the year before. If this is the rate at which headstones are lost, then this historical record will be lost at a much faster rate than originally would have been expected.

Weathering and Stone Types

These are two important factors in the preservation of photography of headstones. Several of the legible headstones in Oxford and York have been photographed; however, for Oxford, there were many that were so weathered that neither date nor name of the deceased were legible. In Greyfriar's kirkyard, Edinburgh this was particularly evident on the headstones that had sandstone and white marble panels for inscriptions; these inscriptions were illegible as the marble surface was nearly entirely weathered. All of the weathered headstones found in churchyards and kirkyards throughout England and Scotland that were not recorded and any information outside of the name, date of death, and possibly age of the deceased that would be found in burial records would be lost.

The key is to preserve the medium in order to photograph it. Inkpen and Jackson (2000) concluded that Oxford contained the most weathered headstones of all six cities that they examined; this was all linked to urban weathering (p. 237). Photographs of such monuments in places like Oxford could help to keep a record of memories that would otherwise be, and have been, lost. According to Mytum (2004a), 'old photographs, such as those [around] 1866 to 1869 and of the 1880s funeral [that] show how some stones had already obtained their patina of lichen whilst others were freshly painted, with design and text picked out apparently in contrasting black and white' (p. 115). The soft stones used in the past will be at risk today; according to Ralston Art Memorial (n.d.), 'it is.possible to get stones that are not made of stone at all but composed from recycled or renewable materials' (para. 3).

Plants on headstones might appear helpful in the preservation of these historical objects, but acids excreted from plant roots only help to speed up their decay. However, painting headstones to preserve the information on the stone, which has been noticed in various places in the UK, and, in all cases, it has helped to preserve the information.

In-Field Issues

Leading to the photograph, in the field there are several elements required to ensure that each photograph is visible and that the photograph does it justice. These elements are, in fact, natural elements such as lighting. Only the photographer will really understand that in order to effectively photograph a headstone or any stone surface containing inscriptions or carvings the angle of the sun (lighting) is greatly important. Although it is possible to manipulate the photograph by using the brightness/contrast feature, this could be attainable in the field with the right lighting conditions. Several gravestone enthusiasts will ensure that bringing a mirror or reflective surface into the field is essential in capturing sunlight; however, in several cases this has not worked in this research, particularly at sites located in places like England and Scotland where days are usually mostly overcast. In this case, it would be better to establish some sort of contrast on the headstone without damaging it. In some cases, some sort of paste or soil was smudged on the face of the headstone. This collects in the engravings and allows the onlooker to decipher them. As long as these methods are non-destructive, they should be used.

Another issue found when working with any kind of monument is connected to measurements. When height measurements have to be taken, it can prove difficult especially with sunken monuments. This can take place because of the decomposition of the coffin, reburials, freeze-thaw action, and other processes or actions that cause the soil to shift and, in turn, the monument also moves. In order to measure the height of the monument, researchers have peeled back the grass to get the full height, if the monument has sunk into the ground. However, it is essential that should this be done, the ground is returned to the state it was found as this can be classified as destructive. Historical monuments, for example, can be quite dangerous as they are very heavy and, if disturbed, can fall. There have been a few cases where people have been killed by falling headstones, or these are a safety hazard (McCarthy, 2009; Park, 2012; Petre, 2006). While in the field, it is important to ensure that all safety measures have been taken. Any form of digging or shifting of material that might be securing the monument could be quite dangerous and should never be executed alone.

All fieldwork done in a churchyard or cemetery should first be requested through the church or cemetery office so that officials know that researchers will be out in the field. In places like a cemetery, staff will drive around as a safety measure.

Selective Preservation

Selective preservation in this case refers to only certain monuments, buildings, and other material culture being kept 'alive' because of their elaborateness and worth to a society or group of people. Those headstones commemorating famous people like Anne Brontë in Scarborough, UK would have been preserved to maintain the memory of such a person and, in turn, her connection to Charlotte or Emily Brontë. This would, therefore, bring money into the community. Although this is positive, it does not promote the preservation of a vast many headstones or other monuments that might possess unique or important elements that could be learned or studied by students and researchers, respectively. This would also represent knowledge lost to all, and, unfortunately, once a monument is erected, it is only a matter of time before this object no longer fits into the social context of that time and, thus, the need to preserve it dwindles.

Although cracked monuments may appear to have lost their appeal to the onlooker, the fact that they have been repaired is part of preservation and is an indication of interest in maintaining that particular stone, either by the family, organisations, or city councils. In other words, a monument does not need to be fully restored to its original vigour to continue to provide the message it might have once been erected to deliver. Yet, had there always been an interest in maintaining historical monuments through recording them (Clarke, 1965), which was common by the mid-1960s, there would be more information available to local schools or groups to learn more about past societies. Instead, those stones that just crumbled were discarded, joining the multitude of destroyed stones near the walls of the churchyard, or if in a cemetery, these discarded stone segments are stored away in an isolated (or hidden) section, away from where visitors could peruse them. Establishing the common practice of photographing historical headstones and adding these to a database is the best code of conduct that could be useful to future generations for genealogical, research, or educational purposes.

PRESERVING A DIGITAL PHOTOGRAPHIC RECORD

When the need for a large number of digital photographs ensues, the most import element is to ensure that all files, right down to the actual photographs, are clearly labelled and stored in known areas. Once a photograph is altered, and is no longer needed, this has to then be updated in all files. However, storing this record in one place can be too risky, as if something goes wrong with the choice of memory storage, and the entire record is eliminated. Besides having the digital record online in a website, these files should all be stored in at least two places to ensure that, if something becomes faulty, it can be easily replaced or is accessible.

Using several USB flash drives could be tedious, but photographs tend not to change much, especially if saved as TIFF. The only thing that might change are the folders in which they remain. In this way, there is not much tedium associated with storing such files. Memory cards for cameras are also useful in storing large files, and for photographs these are essential. USB flash drives and memory cards should also be labelled and stored in one location to ensure that these photographs are not lost.

The case study that follows is an excellent example of how digital photography is useful in fields, such as historical archaeology, where the use of images (pictures) are important in locating trends. Digital photography here has made it possible to capture a trend between motifs and inscriptions on headstones, and this was started long after these photographs were taken in the field. Without these images this study would not have been possible, especially not several years after both the photographs were captured and the monuments were erected.

Case Study: Evidence for the Changing Tradition of Motifs to Epitaphs on Headstones in Some English Churchyards

Abstract: Three cities in England were visited for the purpose of recording headstone information and photographing headstones. Results reveal that Scarborough contains 84% of headstones with both motifs, and 63% with only epitaphs, compared to York (23% and 31%) and Oxford (14% and 6%). Cross-temporal trends are evident, with most motifs and epitaphs dating to the first half of the 19th century, but with the first appearance of motifs in 1608, and epitaphs in 1709. This trend demonstrates the changing levels of literacy in selected areas. One limitation here is the weathering of the headstones, thus, a reduction in legibility (16% with epitaphs). There were dated headstones with illegible epitaphs in Scarborough (57%), York (26%), and Oxford (17%). However, Scarborough contained 75% of dated headstones, 53% for York, and only 26% for Oxford. This indicates that the coastal environment is not weathering the cultural record as much as interior urban sites.

INTRODUCTION

From a historical perspective, motifs have been the most commonly used form of communication on headstones, as the images conveyed understood meanings (Burgess, 1963). Fitzpatrick (1980) enforced this idea by acknowledging that the motifs used had strong symbolism in the community (or parish) that would convey a message without using words, such as the bell and hourglass, which were commonly known for announcing death and the passing of time, respectively. Motifs also benefited those less literate folk in these parishes. After examining the 2000 study on Acton Graveyard headstone inscriptions, it was clear that, like this research, not many headstones contained epitaphs. A representation of only one-tenth of the entire record of headstones, and 11% of those headstones were devoid of information (either through a destroyed headstone or one that was illegible) (PDLHS, 2000). Robb (1998) also mentioned that symbols (or in this case, motifs) have a close relationship 'to power and prestige [which] has become.important' (p. 329). Hijiya's (1983) research examined how the use of motifs enforced the idea of 'variety', which included 'originality, audacity, peculiarity' (p. 355). These were required to attract attention to the stone and, thus, bring focus to the interred and the living family.

As for the epitaph, Hendren (1938) discussed how the advancement of literacy with changing religiosity and taste led to the common practice of using verbal epitaphs. These 'verbal epitaphs' must have been accompanied by images and then to the use of motifs, which were later recorded by those with means. As for the epitaph, its use has been common. In many instances the phraseology of the epithets carved on headstones, particularly in the 18th century in England, are archaic (O'Daly, 1954). This alone is a gateway to the cultural use of the English language that is available to everyone. Epitaphs, as opposed to informative inscriptions, have long been a subject of interest (Willsher, 1985). Voutsaki (2001) mentioned that post-Reformation, the social elite set up elaborate monuments displaying aristocratic lineage through motifs and inscriptions. According to Cannon *et al.* (1989), the higher social classes, with their ability to change most aspects of the society then, had a strong influence on the monument medium, and at an earlier period, tended to use styles prior to their fashionability, and access, to the masses. Crossley (1991) proclaimed that exaggerations in fashion were an expression of status, yet once these fashionable motifs were employed by the masses, their status changed, and these would soon become common, and, thus, not used by the elite. The lower classes would then take these on as status symbols of their own; these lower status individuals had access to styles that had long past their peak in popularity (Cannon *et al.*, 1989). In the US, there was a clear trend in the use of motifs from the 18th century to the 19th century (Dethlefsen & Deetz, 1966). The use of epitaphs on headstones was also awakened in the 19th century in the US (Hendren, 1938). In England, there was a similar trend for both motifs and epitaphs. According to Gittings (2002), this trend was true for several reasons, some being the need to 'display.social status, affective individualism.[and] attitudes to Victorian mourning (p. 121). The use of motifs and epitaphs, but to a greater extent, the latter, could display all of these elements, particularly if churchyards were used quite commonly in the way parks are used today (Burgess, 1963).

Hope (1997) mentioned how the investigation of both text and images are often separated. For instance, in major studies, such as Burgess (1963), Ludwig (1966), Mytum (2000), and Willsher (1985), motifs and epitaphs were explored and relationships between these two aspects, but no trends were established between these on a temporal level. In other studies, such as Dethlefsen and Deetz (1966), motifs are favoured. Hijiya (1983) looked at attitudes displayed through the use of motifs; he also discussed epitaphs, but there is no clear connection of these two elements. In terms of single studies, there tends to be a favouring for one and a lack of connection between the two.

Nevertheless, there has already been considerable work done on motifs. Mytum (2000), Burgess (1963), and Willsher (1985) also addressed motifs and, similarly, divided them in terms of symbolism. Mytum (2000) was very clear on his divisions of motifs: mortality, cherubs and angels, vegetation, crosses, figures, biblical, and marginal decorative elements. Medieval motifs are very similar to those associated with mortality, with the use of scrolls and shields (crests), for example (Burgess, 1963). Willsher (1985) divided motifs into: emblems of mortality, emblems of immortality, carvings of symbolic scenes, emblems of trade, and portraits. Burgess (1963) divided motifs in a similar fashion to epitaphs, except that post-Reformation motifs were grouped into three distinct categories: mortality, resurrection, and means of salvation.

The main purpose of using motifs and epitaphs was to reflect the life of the person interred (Hendren, 1938). Mytum (2000) examined the different types of motifs and epitaphs used to express the loss of a loved one. These sentiments changed throughout history, as mentality to death changed from the Puritan belief to Post-Reformation (Burgess, 1963). Hendren (1938) found that the epitaphs mostly used in the US had a religious affiliation and that the language on headstones tended to become locally idiomorphic. Hall (2001) delved into the idea of epitaphs being linked to the 'culture of mourning' in the mid-18th century (p. 657), and this was particularly relevant in the literary world through poets, such as Wordsworth. Similarly, Sanchez-Eppler (1988) agreed that Wordsworth had an influence on the epitaph: 'an Epitaph must strike with a gleam of pleasure, when the expression is of that kind which carries conviction to the heart at once that the Author was a sincere mourner; and that the Inhabitant of the Grave deserved to be so lamented' (p. 429). Different types of epitaphs have been used, such as biblical quotations, hymns, and poetic verses. Wright (1996) acknowledged that epitaphs reveal the personalities of those who lived before us and mirror the spirit of the times in which they lived, which is a reflection of Deetz and Dethlefsen's (1966) study in the US.

The literature on epitaphs fails to make a clear differentiation between 'epitaph' and 'inscription'. In this chapter, the second-listed author recognises and upholds the division of these two concepts, examining specifically the epitaph. This term 'epitaph' is defined here as the epithet (Mytum, 2000), usually appearing at the base of the headstone, which commemorates the dead through poetic verse, biblical quotation, or other form of commemoration. An 'epitaph' (as established in this chapter) should not include the name of the deceased alongside their date of death or their place of birth or death, as these would be grouped as inscriptions; yet, as most researchers do, an 'epitaph' could be classified under the umbrella term 'inscriptions'. Mytum (2000), Burgess (1963), and Willsher (1985) all made the distinction of referring to epitaphs specifically when discussing epitaphs.

Although Mytum (2000) classified epitaphs under the term 'inscriptions', when referring to introductions as well as biblical quotations, according to him, epitaphs must be further divided into four subgroups: mortality, remembrance, achievement, and religious. Burgess (1963) discussed both inscriptions and epitaphs in one section of his book *English churchyard memorials*, but identified them as separate elements of a headstone. Moreover, these are grouped as Roman, Post-Roman, Anglo-Saxon and Danish, Medieval, and Post Reformation. Willsher (1985) made a clear distinction between inscriptions and epitaphs; she used 'informative inscriptions' to refer to introductions, the name of the deceased, and dates rather than just 'inscriptions', and had a section on epitaphs with detailed examples from parts of Scotland. She focused on popular epitaphs.

The sites of the current study are all located in English cities, and these include Oxford, York, and Scarborough. Two are inland (or interior) study sites and the latter is a coastal site. These English cities were selected on a transect along the UK, linking London to Inverness towards the East Coast. There is no particular (physical or cultural) reason to have selected these cities, but that they were incorporated into a larger investigation in historical archaeology that examined changing trends in headstones for England and Scotland. This research project produced a photographic record of headstones (in their current condition) at six different British cities, including Oxford, York, and Scarborough for England and Edinburgh, Inverness, and Dunbar for Scotland. In total, the record contained 1,088 headstones for England, of which 576 (53%) had legible dates and were considered to be 'legible headstones'. In this chapter, the English churchyards are considered exclusively, as pertains specifically to motifs and epitaphs appearing on the front panels of headstones. The chief aim here is to identify and present any cross-temporal patterns in the use (appearance) of legible headstones, and to establish any potential causalities for these trends. For this reason, a cross-disciplinary approach is taken that also considers weathering of these heritage stoneworks. The objectives are to examine: 1) the use of motifs and epitaphs on headstones in these sites, 2) the transition of motif to epitaph on headstones in selected sites, and 3) weathering patterns on headstones, which can affect the integrity of the record of material culture.

METHODS

As aforementioned, three English cities were examined: Oxford, York, and Scarborough; they were chosen based on their location. Oxford in the centre, York to the north (and a major city outside of London between 1600 and 1900), and Scarborough as a coastal city. The locational sample here allows for the interpretation of different practices towards the use of funerary motifs and epitaphs as is mentioned in various studies (Burgess, 1963; Hendren, 1938; Mytum, 2000).

After selecting the locations, sites were identified based on religious affiliation (Anglican parish churches) of available standing headstones, comprising (a sufficient number of) headstones dating from 1600 to 1902 and a total of at least 100 headstones for each city. Sketch maps were then created for each site with numbers (for later reference) linking an information table to the sketch map. Each legible headstone was photographed using a digital camera. All measurements were recorded on a hand-drawn table. Any imperfections, vandalism, or other damage was also recorded here. The inscriptions (introduction, name, and date of death) were recorded. Burgess's (1963) method of categorising motifs and Mytum's (2000) method of categorising epitaphs were exploited, as they both include three categories. Burgess's (1963) method (for motifs) included: 1) mortality, 2) resurrection, and 3) salvation; and Mytum's (2000) method (for epitaphs) includes: 1) mortality, 2) salvation, and 3) remembrance. The motifs and epitaphs used in this study were, thus, divided and counts were made to find the different proportion for each city and for all of England.

RESULTS

Oxford contained four sites, York three, and Scarborough one (Table **1**). Table **2** shows the number of headstones from the total record for the three cities that were classified as illegible because they were obstructed with a thick ivy cover or weathered. There were 25 (27%) headstones with motifs and 69 (73%) without motifs. Of the 94 headstones, 2 (2%) were dated to the 17th century, 13 (14%) to the 18th century, and 79 (84%) to the 19th century.

Table 1: Churchyards within the selection of English cities in this study

City	Churchyard
Oxford	St Peter-to-the-East
	St Giles
	St Mary Magdalen
	St Cross
York	Holy Trinity
	St Olave's
	St Denys
Scarborough	St Mary's

Table 2: Summary of the appearance of motifs and epitaphs on dated headstones

Total for English Sites	Motifs	Epitaphs	Both	Neither
First Appearance	1608	1709	1783	1708
Last Appearance	1889	1868	1867	1895
Peak Appearance	1810-1819	1830-1839	1820-1829; 1850-1859	1830-1839
Duration of Popularity	281	159	84	187
Total Headstones (507)	*88*	*87*	*27*	*305*

*69 headstones were excluded due to their illegibility

Table **3** illustrates the first and last appearance of motifs and epitaphs on headstones for all sites, the peak appearance of these, their duration of popularity for England, and the total numbers containing motifs and epitaphs. The table shows that motifs were the first to appear (of the English sites examined) and that epitaphs followed a century later. The first instance of the use of both motifs and epitaphs on a headstone was in the late 18th century, 80 years after the first appearance of an epitaph. However, there is a parallel in the appearance of epitaphs and the lack of use of either motifs or epitaphs on headstones in the early 18th century. Moreover, the use of motifs on headstone lasted for nearly 300 years, which is just double that for epitaphs and even less for neither appearance. The use of both motifs and epitaphs lasted for less than a century. As for the proportion of headstones in each city, Scarborough contained the majority of headstones (59%), followed by York (24%), and Oxford (17%). Each site contained illegible headstones (those that did not have visible dates) and, although these were mapped, they were not used in the research. In fact, these sites contained 47% of illegible headstones. Of these, Oxford contained the highest proportion (74%), followed by York (47%), and Scarborough (25%).

Table 3: Summary of the appearance and absence of motifs on headstones with illegible epitaphs

Motifs/Century	17th	18th	19th	*Total*
Legible	2	3	20	*25*
Illegible	0	10	59	*69*
Total	*2*	*13*	*79*	*94*

Scarborough contained more headstones with legible epitaphs (63%) as well as those with illegible epitaphs (57%). York followed with 31% of headstones with legible epitaphs and 26% of those with illegible ones. Oxford contained the least number of headstones with epitaphs (6%) and a similar amount for those with illegible epitaphs (17%). Of the headstones with motifs, 64% were from Scarborough, 23% from York, and 14% from Oxford. Very few headstones contained both motifs and epitaphs; Scarborough contained the most (57%) of these, whereas York and Oxford contained 22%, and 21% of headstones with no motifs and epitaphs.

Table **4a** summaries the numbers of headstones with motifs within each site. Oxford contains 83% of headstones from the 19th century, with an equal proportion from both the 17th and 18th centuries. York also contains a large proportion of headstones, with motifs dating to the 19th century (65%), but 35% are from the 18th century and none from the 17th century. For Scarborough, 84% are from the 19th century, 14% from the 18th century, and 2% from the 17th century. Similarly, Table **4b** shows that this pattern persists for epitaphs, with a larger proportion of headstones dating to the 19th century (Oxford 100%, York 87%, Scarborough 80%) and the previous centuries containing less than 50% of headstones with epitaphs (18th century: Oxford 0%, York 19%, Scarborough 20%; 17th century containing no headstones with epitaphs).

Table 4a: Summary of the appearance of motifs on headstones by century within each city

City/Century	17th	18th	19th	Total
Oxford	1	1	10	12
York	0	7	13	20
Scarborough	1	8	47	56
Total	2	16	70	88

Table 4b: Summary of the appearance of epitaphs on headstones by century within each city

City/Century	17th	18th	19th	Total
Oxford	0	0	5	5
York	0	5	22	27
Scarborough	0	11	44	55
Total	0	16	71	87

Willsher's (1985) method of categorising motifs and epitaphs was not employed in this study, as the categories that she included were more suited to Scotland rather than England. Hence, in a future study of Scottish kirkyards, her method will be employed.

Table **5a** illustrates that there were 115 headstones sampled in the three English cities; the first two categories of motifs were not extremely evident, with only 2 (2%) of headstones containing motifs of mortality, and 4 (3%) of those with motifs of resurrection. Of salvation motifs, 109 (95%) were represented for England. Oxford and Scarborough both had 1 (50%) headstone with motifs of mortality (skull, hourglass), but York contained no such motifs. For motifs of resurrection, Oxford contained no motifs, whereas Scarborough had 3 (75%) of these and York had 1 headstone with such motifs (cherubs with hourglass, candles). For motifs of salvation, all cities contained headstones with these types of motifs; Scarborough represented 69 (63%), York 25 (23%), and Oxford 15 (14%). These included the cross and urn as major motifs. Table **5b** illustrates that, from the three cities sampled, 114 headstones contained epitaphs: 72 were from Scarborough, 33 from York, and 9 from Oxford. Similar to motifs, epitaphs of mortality were few; a total of 9 (7% of all epitaphs) were found and the majority were from Scarborough (6), followed by York (2), and Oxford (1). Salvation epitaphs were the most popular, with a total of 61 (54% of all epitaphs), where 40 were from Scarborough, 15 from York, and the remaining 6 were from Oxford. Remembrance epitaphs totalled 44 (39% of all epitaphs); Scarborough contained the most (26), followed by York (16), and then Oxford (2).

Table 5a: Summary of the types of motifs found on headstones in each churchyard according to Burgess (1963)

City/Category	Mortality	Resurrection	Salvation	*Total*
Oxford	1	0	15	*16*
York	0	1	25	*26*
Scarborough	1	3	69	*73*
Total	*2*	*4*	*109*	*115*

Table 5b: Summary of the types of epitaphs found on headstones in each churchyard according to Mytum (2000)

City/Category	Mortality	Salvation	Remembrance	*Total*
Oxford	1	6	2	*9*
York	2	15	16	*33*
Scarborough	6	40	26	72
Total	*9*	*61*	*44*	*114*

Fig. **1** shows the variations of motifs on headstones in all three cities. For unrepeated motifs, 40% were relating to mortality, 14% to resurrection, and 73% to salvation. For repeated motifs, 60% were relating to mortality, 86% to resurrection, and 27% to salvation. Overall, 8% were connected to mortality, 11% to resurrection, and 80% to salvation; 64% of all motifs were unrepeated, while 36% were repeated; for example, from the 1600s to 1700s and then to the 1800s.

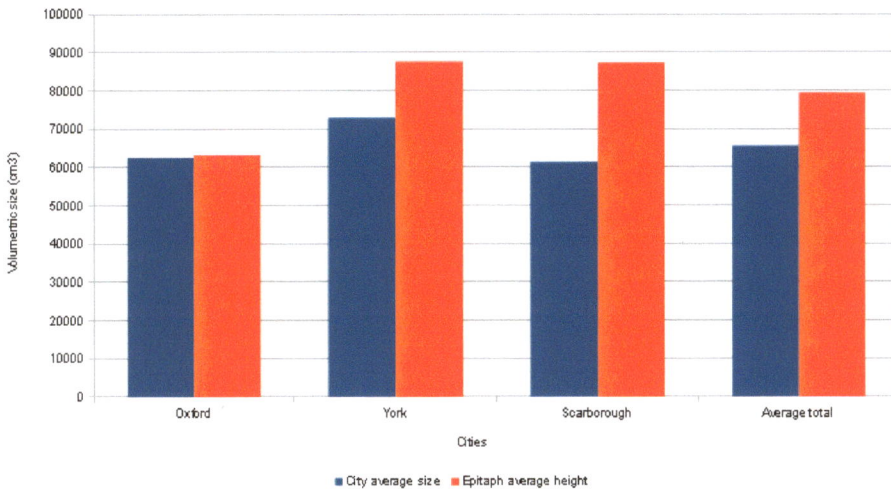

Figure 1: Summary of the average size of headstones in each city and the average size of headstones with lengthy epitaphs.

Fig. **2** shows the size of headstones during the Victorian period; the average headstone (volumetric) size was 65,585 cm^3 for all three cities, while the average for headstones with lengthy epitaphs, this was nearly 14,000 cm^3 bigger. In Oxford, the headstone average was 5% smaller than the city average, for York this was 11% smaller, and for Scarborough this figure was as low as 6% smaller. Comparing the city headstones with those housing lengthy epitaphs, Oxford's average was 20% smaller, while York and Scarborough were half the size, with 10% smaller headstones.

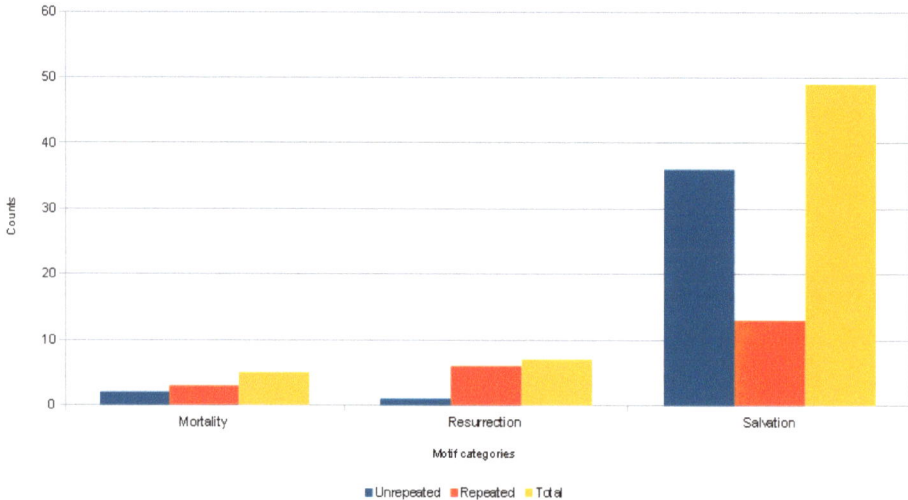

Figure 2: Summary of the categories of motifs, either unrepeated or repeated, for all three cities.

Some motifs, such as floral, were repeated throughout the three centuries (1600s to 1800s); others, such as radiance, sun, and urn, were popular in both the 1700s and 1800s. In Oxford, the scroll motif was used in the 1600s and then in the 1800s, while for Scarborough it appeared from 1720. Also, in Scarborough the draped urn appeared around 1600s to 1720 and then from 1781. In York, the spade weapon was unique to 1720-1780, and in Scarborough the instrument motif was unique to the period from 1781.

DISCUSSION

The results show peaks in use of either motifs and epitaphs. As Cannon *et al.* (1989) mentioned, the wealthier part of society had more control over the fashion, which meant that they used styles before they became common fashion. He also stated that those with less means made use of fashionable materials once they were no longer popular. If this were the case with epitaphs, and they were introduced in parts of England in the early 18th century, it could suggest that these headstones were more likely to belong to the wealthier classes, inferring that subsequent peak in popularity, of either epitaphs or motifs, might have been caused (to a greater degree) by those individuals of lesser affluence. Meyer (1990) addressed the issue of how epitaphs were also used to convey familial relationships as well as status, which led to the growth of epitaphs.

Mortality motifs are classified as those that appeared before the Georgian period and up to the beginning of it and the dates assigned to this category were 1600 to 1719. These motifs were connected to the concept of 'death' and 'time' (Meyer, 1990, p. 165). These usually included a primary, or significant, motif, such as a skull, with additional motifs, such as bones, sexton tools, hourglass, sundial, and candle (Meyer, 1990, p. 166). Resurrection motifs are a significant element of the 18th century, and the specific time period chosen for these was 1720 to 1780. These had some emphasis on death in the earlier years, but this diverts to focus more on heaven; thus, usually involving major motifs, such as angels or cherubs, coupled with motifs symbolising death, time, and eternity (Meyer, 1990, p. 166). These motifs eventually led to the use of the urn up to the end of the Georgian period 'as being the most conspicuous item in this pagan repertoire of ornament' (Meyer, 1990, p. 166). Salvation motifs represent the last half century of the Georgian period and the chosen dates for this was 1781 to 1902. During this time, faith, hope, and charity replaced death and time (Meyer, 1990, p. 166). Biblical scenes are also included and, with religion and the Gothic idiom in architecture, the beginning of the Victorian period bring motifs, such as the cross (Meyer, 1990, pp. 166-167).

The majority of the headstones sampled were dated to the 19th century, so to have such a large population representing salvation motifs is not surprising. One aspect of the finds that was surprising, however, was the nil sample for resurrection motifs in Oxford. This could have been due to the small sample size for this city, irrespective of the number of churchyards sampled. Moreover, the nil sample for mortality motifs for York was also surprising, as it would have been expected that in York, being a populated city in the Georgian period, that there would have been more Puritan influence, or the evidence of this on headstones. However, considering St Denys's history, as one of the churchyards sampled in York, it is not surprising that a limited variety of motifs were found. According to Kightly (2004), the church as it stands 'is only the east end of the Mediaeval church, which was originally three or four times bigger' (p. 3). If this is the case, that means that 66% to 75% of the possible headstones from this churchyard were destroyed in the Civil War Siege of York in 1644 and during the collapse of the west wall in 1797 (Kightly, 2004).

Mytum's (2000) use of different epitaph types is similar to Burgess's (1963) motif categories. These types are: 1) mortality, 2) salvation, and 3) remembrance. Mortality emphasised the presence of death, specifically through the corpse, or the existence of a body (Mytum, 2000). These epitaphs were meant to display 'Human frailty, physical and spiritual, [as] highlighted in [the] texts which [served] a particular purpose in communicating the way to salvation which could still be taken by those alive' (Mytum, 2000, p. 51); Mytum (2000, p. 51) also contains:

> Remember, Man as you Pass by,
>
> As you are now so once was I:
>
> As I am now so you must be
>
> Therefore Prepare to follow me.

The second category, salvation, usually focused on immortality, that death would bring a better life and that there was 'compensation in the knowledge that the deceased [had] passed to a better place' (Mytum, 2000, p. 53). Mytum's (2000, p. 52) example of this is:

> Gentle reader, let your prayer
>
> Be for our souls and that sincere,
>
> Lord have mercy on us thrice say
>
> Eternal rest enjoy we may

The final category, remembrance, came to be with the decline in salvation where the dead were recalled with affection and respect. These included epitaphs of 'occupations, residence, and achievements or qualities' as well as a 'wider advertising of social and economic status on [the] public monument' (Mytum, 2000, p. 54); for example (p. 53):

Husband and father one of the best

Jesus calls thee to thy rest

According to the evidence for epitaphs, York's evidence of mortality (even in epitaphs) is quite low, which displays strong evidence towards loss of culture through past destruction. Similarly, St Mary's church in Scarborough only contained 8% of headstones of mortality, which would have been popular in the 1600s to early 1700s. Between 1400 and 1600, St Mary's church was threatened and parts destroyed during six sieges against the castle (WSMCS, 2007).

Any gaps in this record is either due to the management of the churchyards, vandalism, or illegibility (and, thus, removal), which has consequently weakened the results. Moreover, based on the literature, mortality motifs, popular before the Georgian period, are not strongly represented in the results due to this gap in the record. Nevertheless, the use of epitaphs appears to have increased in use in the 19th century, compared with the previous two centuries in the US, and this is also evident in the results, where 82% of the headstones dated to the 19th century had epitaphs compared with 18% in the previous century. Yet, the results also show that, unlike in the US, in England, epitaphs were introduced as early as 1709 and motifs a century earlier in 1608.

In some cases, it is not whether an epitaph was used or not, but whether the record is able to capture their appearance in its current state of preservation. For this reason, it is necessary to consider the physical condition of the cultural record in addition to its cultural expression, as this may indeed affect the interpretation of the archaeological record. Dissolution (granular disintegration) due to acidic rainfall (Fig. **3a**), for instance, has been observed to deteriorate the condition of headstones in polluted urban environments (*e.g.*, Hoke & Turcotte, 2004; Lammel & Metzig, 1997; McNeill, 1999; Reddy, 1988). The dissolution process is additionally affected by the presence of biofilm microbiota (Mitchell & Gu, 2000), which is evident in the study locations predominantly as algae and lichens. More recently, Thornbush (2012) developed a weathering scale for limestone that has been applied to headstones found in central Oxford (Thornbush & Thornbush, 2013). Their study illustrates that complexity of the weathering system, where pollution and microclimate (urban climatology, *e.g.* Camuffo *et al.*, 1999) contribute to the formation of weathering features.

a)

b)

Fig. 3: contd…

c)

d)

Fig. 3: contd…

e)

f)

Fig. 3: contd...

g)

h)

Fig. 3: contd…

i)

Figure 3: Weathering pattern affecting the preservation of epitaphs on headstones, including: a) dissolution at Oxford, b) encrustation (alveolar) Scarborough, c) exfoliation along bedding planes of limestone at Oxford, d) evidence of salt weathering at Scarborough, e) disappearance of the record towards the middle to bottom of the headstones at Scarborough, f) blistering and bursting erasing the record at Scarborough, g) honeycomb (alveolar) and cavernous weathering at York, h) burial at Scarborough, and i) placement of epitaphs and the impact of weathering at York.

Even though some headstones in Oxford are completely clear of epitaphs (and inscriptions in some cases), this may not be the reason for their exclusion in this city. Some headstones appear to have space on the front panel for (at least short) epitaphs, but they are missing (Fig. **3b**). Here, it is evident that blistering can impact headstones once crusts have formed (due to exposure to urban pollutants, such as sulphur from coal combustion), ruining the record through the development of blister domes that burst and exfoliate (Viles, 1993). This is evident in Oxford, where the water table is high due to its location in the Thames valley and salts dissolved in groundwater are able to move up headstones through the 'wick effect' (Goudie, 1986), effectively removing details towards the bottom of headstones, where epitaphs would be situated (Fig. **3c**). Salt weathering is also clear at Scarborough because of its coastal location (Fig. **3d**). Here, salts have migrated further upwards in the headstone, causing illegibility higher up towards the middle of headstones (Fig. **3e**), where crusts have exfoliated (Fig. **3f**), and in some cases honeycomb (alveolar) and cavernous weathering are evident (Fig. **3g**), produced through the deposition, ingress, and evaporation of saltwater (Mustoe, 1982; Young, 1987).

The height of the headstone above ground-level is affected by burial. In some cases, headstones are buried deeply, hiding most inscriptions and epitaphs (Fig. **3h**). Elsewhere, however, there is sufficient space remaining on headstones so that burial does not in the least affect the appearance (and legibility) of epitaphs (Fig. **3i**). In addition to environmental factors affecting the preservation of the record, it is important to note differences in rock type, which may trigger certain types of weathering, such as sheet exfoliation (contour scaling) along bedding planes on limestone. Softer rocks, such as those containing more sand granules (layer particles), are susceptible to cavernous weathering at the coast, as is clearly evident at Scarborough. At this coastal site, the inheritance of past pollution conditions (the 'memory effect' observed for sandstone by McCabe *et al.* (2007) is at work in the form of exfoliating black (sulphurous) crusts. All of this natural decay is additional to the destruction of the cultural record due to warring, as for York.

According to Snell (2003), Victorian headstones were large enough to house 'lengthy epitaphs and biblical texts' (p. 115). The English sites examined contained headstones of varying sizes; Snell's concept of the Victorians' manufacturing of large headstones does seem to parallel the evidence gathered in this study. Moreover, there were some headstones that were of considerable size that were devoid of epitaphs; this shows that although headstones were being manufactured larger, this was not for the sole purpose of housing epitaphs. However, during the reign of George III in Scarborough and York, lengthy epitaphs were present, irrespective of the size of the headstones. In any case, places like Oxford contained headstones that did not really vary in size nor contained lengthy epitaphs. Nevertheless, headstone size is supported in Scarborough, where headstones during the Victorian period were bigger in size and contained epitaphs, so that smaller headstones were devoid of epitaphs. The use of epitaphs on headstones in this study peaked in the 1830s, as based on the evidence.

Mytum (1994) identified how those in England who were literate had a choice of phraseology in order to denote their intelligibility as well as social standing. Vovelle (1980) noted that in the 17th century epitaphs were used by the urban and rural elite. However, up until 1740, only great notables were worthy of an epitaph. It is difficult to decipher which headstone belonged to the elite or not, as some of the poor also erected headstones to commemorate their dead, and the use of epitaphs increased amongst most classes by the end of the century. Nevertheless, Wright (1996) believed that throughout the Middle Ages it was only the wealthy who had memorials, and according to *Woodward Family Tree* (2013), poor families could not afford to purchase a headstone, and infants would be buried with an adult to avoid the cost of one.

In the latter part of the 19th century, much of the verse and inscription on churchyard memorials were concerned with mortality; its inevitability and its levelling powers (Wright, 1996). Mortality epitaphs found in this research were more prominent in Scarborough than in the other two cities, and these were particularly popular in the 18th century. For example, at Scarborough in 1790:

> For all flesh is as grass, and all her glory of man as the flower of grass.

> The grass withereth and the flowers thereof falleth away. But the word of the Lord endureth forever.

> A Truth, O Friend, which you must shortly prove!

> Then, haste to see substantial Joys above.

> Fondly I dreamt of Happiness below.

> But quickly found earth's pleasures end in woe.

> The beams of Grace chased darkness from my eyes.

> Heaven, through Afflictions, made the Sinner wise.

> Happy for me. Life's dangerous Voyage is over.

> Nor Storms nor Griefs, on high

> Art thou not dead, yes and here I live.

> I who on Earth, men did live to die

> Died for to live with Christ, Eternally[*]

*Spelling mistakes on the headstone

Such epitaphs contained an air of darkness and crude treatment of life as an existence. Which was also evident at Scarborough in 1709:

All you., my Grave to.

Weep for [thy] selves and not for me.

.in my.was snatched away,

Therefore.make no Delay.

However, in the same city, in the mid-19th century, a similar epitaph of mortality contained less darkness, or simply less detail about the corpse and death itself, as at Scarborough in 1832:

My [life] was short; my years had few

Like us the grass.

.behind.

In Oxford, there was only one headstone with a legible epitaph of mortality (from 1868):

My flesh shall slumber in the ground, till the last trumpet's joyful sound then burst the chains with sweet surpr and in our saviour's image rise.

This study reveals that epitaphs referring to salvation were more popular, such as for Oxford in 1866:

Into thy hands I commend my spirit;

for thous* hast redeemed me. O Lord,

thou God of truth

*Spelling mistakes on the headstone

In York and Scarborough, a popular epitaph containing qualities of salvation was:

Blessed are the dead which die in the Lord.

This epitaph appeared around early to mid-19th century. Remembrance epitaphs followed with 39% and the recurring epitaph in Scarborough was:

In life respected, in death lamented.

This epitaph was used mostly in the mid- to late-19th century. In this research, more epitaphs were used in the first half of the 19th century in comparison with the previous centuries. They appeared after motifs, which are prevalent in earlier times, when the literacy rate was low (as in the Middle Ages).

Crossley (1991) dedicated a chapter to epitaphs, and described how the English churchyard contains 'quaint, cryptic [and] amusing epitaphs' (p. 89). Based on Hendren's (1938) idea that religion played a role in epitaphs and that, with time, they became more idomorphic, might be true for these sites, as some were unique to the place; but, the lack of an epitaph can also be a form of language, with the disuse of one for reasons known or unknown.

The link between motifs and epitaphs is clear; in the mortality period during the 1600s to early 1700s, while skulls and cloths were used, dreary epitaphs with emphasis on the corpse were prevalent. It is clear that during this period of time (1600-1719), the message being conveyed was of death and to prepare for this. During the resurrection period (for motifs) and salvation period (for epitaphs) (1720-1780), sun and radiance motifs began to be used and epitaphs were more focused on freedom from pain and more on hope that the deceased would reach heaven. This focus on the soul is clear in the appearance of radiance (as is the idea of hope). The use of cherubs generally began to be found during this time, but not in these cities. The use of the cherub motif became popular during the period of salvation and, during this time, epitaphs were in the period of remembrance. The use of the cross and urn became more pronounced in these cities, as did candles and the broken dart. As for epitaphs, these were meant to be more focused on the individual and their successes in their life as with: 'In life respected, in death lamented'.

CONCLUSION

The research has shown that the cultural record for inland polluted sites can be worse off than at the coast, where headstones are exposed to marine salts and wind. For the selected interior sites of Oxford and York, sulphurous crust formation, blistering, and exfoliation (bursting) can do away with inscriptions, and with them epitaphs, which are usually located towards the bottom on the front panels of English headstones. These are most affected, however, by rising damp (from capillary rise of groundwater) migrating up headstones and dissolving as well as weakening rock fabric, leading to its exfoliation. The severity of exfoliation is impounded by salts at the coast entering the rock fabric forced in by wind action, leading to cavernous forms of weathering.

Exposure and the condition of the headstones does affect the legibility of headstones, with epitaphs being absent from a majority of headstones where motifs were located. However, this could be a cultural product, as motifs led the way to epitaphs as a preferred form of cultural expression on headstones commencing from the late 18th century and leading well into the mid-19th century, showing a disparity from 1820s to 1830s, whereby the use of epitaphs nearly doubled that of motifs, but continuing into the 1870s, when this trend disappeared. Although this trend is quite evident in these findings, a majority of headstones recorded were devoid of both motifs and epitaphs, which peaked in the 1830s. In any case, there is a clear seriation in this study, with the peak of motifs in 1810s, leading to the peak of the use of both motifs and epitaphs in 1820s, to the peak of only epitaphs (and headstones without either) in the 1830s, to the return of the use of both in the 1850s. In the 1840s, more headstones had epitaphs rather than motifs, but the majority were devoid of these.

ACKNOWLEDGEMENTS

We are grateful to the porters of St Edmund Hall at the University of Oxford for their hospitality. Also, the rector at St Olave's church in York for access and support. We are also grateful to those who took the time out to meet us to support us in our fieldwork; and a special thank-you goes out to the priest at St Denys's church, York and the priest at Old High kirk, Inverness. Gratitude for the provision of materials goes out to the priest at St Giles's church, Oxford; the wardens of St Mary's church in Scarborough; the helpful assistance at Greyfriar's, Canongate, and St Cuthbert's kirks in Edinburgh, as well as the priest at Dunbar Parish church. Access was provided into all churchyards/kirkyards and we thank the various people involved for their assistance, both in England and Scotland.

Chapter 8: Heritage Sustainability

CHAPTER 8

Heritage Sustainability

Abstract: Cultural heritage has continued to be explored by several organisations in the UK. Understanding heritage is important in learning about current society, and, in several ways, the churchyard is a strong source for learning about cultural history. The changing styles of headstone inscriptions provide information about the socioeconomics of a place; and regional styles provide information on the fashion of the time and the understanding of material culture.

Keywords: Churchyards, conservation, headstones, inscriptions, kirkyards, material culture.

What is the importance of sustaining heritage? Several different organisations that specialise in this believe that it is essential for various reasons, and some of which are reasons probably, to the masses, not associated with heritage. According to Costa and Batista (2014), places of heritage have a disperse history; these sorts of places could provide ample information about a people. English Heritage looks into conserving and preserving history as well to learn more about a England and what creations paved the way for current society. However, according to Historic Scotland (n.d.), there are several reasons why heritage is important, some of which include enhancing our quality of life (para. 2) and environmental regeneration (para. 6). These are two very good points as they are both linked with well-being and appreciation of local heritage of which something like a historic building, such as church, or even churchyard, could provide.

The sustainability of heritage in the UK is examined here in part through the perspective of the churchyard. Headstone examination has been employed in various facets of research to include archaeological, educational, English language, demographics, genealogical, and social studies. The necropolis is an ideal time-capsule, whereby a vast amount of information is available to the onlooker. Despite being offered in various different perspectives, this information is a stable form of learning about ancestors, gentry, and the populace who lived many decades before. Moreover, the fashionability of the past can be compared to that of the present, which allows for more knowledge transfer through this medium. Walking along a churchyard or a burial ground, allows one to explore the endless possibilities of commemoration.

The way the dead are remembered is very much revered and is a part of culture. These beliefs are vividly exposed on a headstone to show how much the dead were loved whilst still alive. The carvers in the past played a very unique role in preserving the heritage that can still be seen on headstones in contemporary society. These elements are not only associated with language, but with art and logic. In fact, headstones through time have not changed drastically. The signposting of introductions, the name of the deceased, the place of birth or death, and all other inscriptions are all carved in a particular font, unique in size and style. Examining different churchyards and a vast array of headstones, *in situ* and in museums, confirms that there was a particular order, and to some degree this particularity has seeped into the modern headstone.

Heritage conservation is a stable area where research continues to develop. According to Amoeda *et al.* (2012) various elements are involved in these new approaches such as pollution, poverty, religion and war to name but a few (p. V). Heritage conservation can ultimately resolve these types of issues by generating money through tourism and the lottery fund to help support the maintenance of historical monuments. However, another important way of helping maintain such cultural heritage is to create policies against the destruction of historical headstones. The aim of the current headstone research is to sustain a part of the UK's history through photography. Many government organisations aim to work on headstones directly by restoring those that have been grafittied, broken, or are victim to weathering. The latter is the basis for going into photographic research; through starting a database for historical headstones, these could be available for more of a duration, and they could continue to be accessible to the masses. This plays an important role in genealogical research as well as education.

Currently, there are several platforms that allow for the storage of photographs, and these have been used more on a personal level than for research. What this means is that there are several different sites that one could visit and this could lead to having to surf the web in an attempt to seek out a name. Yet, there are genealogical sites that provide the information needed, but many of these require a fee, which could be rather expensive for those people who do this on a one-time basis. Generally, to have all photographs housed in one location to be accessed by anyone who is interested is the aim here.

Ultimately, as these headstones are situated outside in churchyards within urban areas, they are directly affected by the environment and vandalism. All of the churchyards in this research have shown signs of both, and although the churches and city councils attempt to preserve or clean these monuments, it is only time before they have to revisit them. In this way, photographing and documenting each headstone into a database would be one step towards preserving this material culture and, thus, sustaining an important part of Britain's heritage.

The case study that follows illustrates how cultural heritage could be investigated through the use of headstone inscriptions in the UK. Analysing inscription trends sheds light on language use, beliefs, and craftsmanship in different regions in history. In fact, several researchers have conducted in-depth analyses of headstone inscriptions in different parts of the UK to get a better perspective on cultural change and some of these will be addressed in this study.

Case Study: The Changing Styles of Inscriptions on Headstones in English Urban Churchyards and Scottish Urban Kirkyards

Abstract: This study examined the varying styles of inscriptions on headstones in six different cities within the UK. Elements such as introductions, names of the deceased, place of death, and other were examined with social, traditional, and economic implications attached to these choices. The reasoning behind incorporating varying font styles and sizes on one headstone have social implications the include headstone size, literacy rate, and choice of phraseology. Traditional reasons include familial expectation and religion, while economic reasons involve industry and wealth. The headstones date from early 17th century to the end of the Victorian period. Generally, a majority of headstones included at three font types, whilst in England, four styles were found in the coastal site of Scarborough. Scarborough contained 58% of the total headstones for England, while York followed with 24% and 18% for Oxford. Of the Scottish sites, Edinburgh comprised 55% of the total headstones for Scotland, while 28% came from Dunbar and 15% from Inverness. The differing font styles include Gothic, Roman, Italics, Script, and Fishtail and these were chosen for the different elements such as Gothic for introductions, Roman for upper case names and most inscriptions, Script for place, and Fishtail for introductions.

INTRODUCTION

Studies about headstone inscriptions have been written in several different scales. Burgess (1963) and Willsher (1985) compiled studies looking at several burial monument details in burial grounds (cemeteries and churchyards and kirkyards) to include inscriptions as well as other features. Similarly, Mytum (2000) briefly wrote about the varying inscription styles amongst the detailed description and analysis of other monument features in cemeteries around England. Brown's (2008) study focused particularly on motifs and symbolism, but identified some elements of inscriptions, without deviating from their meaning, rather than just focusing on the morphology of these features. Strangstad (1995) explored, although minimally, inscriptions in an educational, weathering, and general manner of use such as in education and cultural studies. These studies are important in the understanding of the fundamentals of inscriptions and their meanings and productions within burial grounds as a representation of once-living members of society. There have been other studies that ventured closer at looking exclusively at the study of inscriptions. Wright (1996) looked specifically at epitaphs, identifying the different significances in meaning of these, creating distinct divisions of these. In this work, he dedicated an entire chapter to inscriptions and epitaphs, taking note of the differing types of inscriptions and font styles. There have also been other studies that simply recorded the inscriptions on burial monuments within burial grounds. These studies are relevant in comparing names and dates to fieldwork that has already been conducted and could be used as a primary resource for future studies. Brown's (2012) work contained inscriptions found on monuments in Greyfriar's kirkyard in Edinburgh; and there have been several studies similar to this one.

The study of inscriptions is useful in understanding historical trends relating to death, which in turn, provides a basis for the comprehension of how people in the past perceived and dealt with death. Over time, changes have taken place of the types of inscriptions and their purpose on the headstone. Wright (1996) observed that inscriptions in the 1600s tended to be 'brief and factual', but later in the 1700s and 1800s these grew in length, becoming more 'verbose, sometimes ingenious, often vainglorious, tender and brutal, sublime or ridiculous and.witty' (p. 19).

This research examines inscriptions in specific English and Scottish sites by focusing on the number of varying styles that have been used and why, to some degree, these have been chosen. According to Crossley (1991), churchyards in Britain exemplify art and skill that once existed (p. 11). Brears (1981) mentioned that headstones are useful in studying regional styles of decoration (p. 84). This study aims to examine the differences in inscriptional styles within selected cities in Scotland and England. The English sites include Oxford, York, and Scarborough. The former containing four churchyards: St Mary Magdalen, St Giles, St Peter-in-the-East, and St Cross. The churchyards in York include those belonging to St Olave's, St Denys, and Holy Trinity. For Scarborough, there was only one churchyard examined, St Mary's. The Scottish sites include Dunbar, Edinburgh, and Inverness. The former and latter sites contain only one church, Dunbar Parish and Old High, respectively, while Edinburgh contains three sites: Greyfriar's, Canongate, and St Cuthbert's.

These cities have varying histories; of the English sites, however, they were constructed between the 10th and 12th centuries, with restorations conducted between 12th and 19th centuries. These were also commonly disused for burials in the late 19th century, with St Peter-to-the-East, one of the Oxford sites, being disused in the 20th century. The York sites were generally built in the 11th or 12th centuries and restored in the 13th to 15th centuries. These sites were disused for burials in the late 19th century. Scarborough was constructed in and around the 13th century and restored in various periods up until the 19th century. It was disused in the late 19th century.

The Scottish kirks were founded much later, the earliest being in 17th century. However, these refer to the actual buildings, while in some cases, there might be references to an actual kirk existing in historical records. Dunbar Parish was built in the early 19th century, but rebuilt after the late 1980s from a fire that destroyed most of the edifice, alongside important monuments that might have been housed within the church. In Edinburgh, the older kirks belong to Greyfriars and Canongate, having been built in the 17th century. St Cuthbert's was founded around a couple of hundred years later. However, St Cuthbert's and Greyfriar's kirkyards were disused in the late 19th century, while, according to the church, Canongate can still house burials as long as they are contracted to do so. In Inverness, St Stephen's (Old High), was founded in the late 19th century.

This comparative study examining the number of inscription styles used on a given headstone at a national scale provides a cross-cultural study for Scotland and England. This brings into view the socioeconomic influences associated with the choice of inscriptions, which provide some basis into the general style of the period that influenced art and architecture, of which both are evident in the graveyard. From here, the understanding of the use of specific styles, to include fonts, comes into play. Regional styles, as mentioned by Brears (1981) shed light on the people and the fashion at the time, to provide an understanding of the material culture. Moreover, as mentioned by Mytum (2004a), the inscriptions on a headstone provide a story of the background of both the interred and the period of time in which the stone was constructed (p. 113). The aim here is to find the differences in number of inscriptions used on individual headstones in both England and Scotland.

METHODOLOGY

Fieldwork was conducted at the English and Scottish sites at different times between 2006 and 2013. Within each site, a map was manually constructed to contain headstones *in situ* in relation to the church. Headstones were photographed using a Nikon S4 digital camera. They were subsequently measured with a tape measure and Cardinal readings were taken using a Silva compass. A manually-constructed table was created containing different fields such as: the number of the headstone, which was to be located on the map of the site, the date of death, the measurements of the headstone (width, thickness, and height), the type of stone, the headstone shape (in numerical classification according to Mytum, 2000), the Cardinal reading, and any additional details that were derived from each headstone. The latter usually included whether there were motifs, the type of introduction used, whether there was an incline (or whether the headstone was tilted), any weathering features that affected the headstone quality, the name of the deceased, the place of death, and whether there was an epitaph. These characteristics were important to be documented on-site as they helped with the initial understanding of any immediate patterns that could be derived from individual sites, and which could be, thus, compared to other sites. However, the majority of these data are relevant to a larger study.

The current study was derived from a much larger study that looks at all aspects of headstones in six different cities in Scotland and England. For this particular study, counts were used from each site; these were grouped in one, two, three and four inscription fonts found on each headstone. Patterns were derived from the Scottish and English tables A χ^2 test was then conducted to decipher the statistical significance of the data for all English and Scottish sites.

RESULTS

The proportions of the varying font styles in both Scotland and England are presented in Tables **1a** and **1b**, respectively. These tables provide clear patterns based on the number of inscriptions used on the different headstones encountered at each site. Of the Scottish sites (Table **1a**), there is a clear pattern in the dominant use of one font type; half of this amount is evident in the use of two fonts and a much less proportion for that of three fonts used on a given headstone. The use of four fonts was not evident in the Scottish sites.

Table 1a: The number of different fonts found on individual headstones in the Scottish kirkyards examined

Scotland	One Font	Two Fonts	Three Fonts	Four Fonts
Dunbar	64.57%	31.89%	3.54%	0.00%
Edinburgh	66.46%	31.03%	2.52%	0.00%
Inverness	61.64%	34.93%	3.42%	0.00%

The English sites (Table **1b**) have a more scattered pattern, whereby there was no clear progressive pattern detected. Oxford headstones contained more one fonts, while York and Scarborough contained more two fonts on the headstones. However, the use of two fonts in Oxford was quite comparable to that for one font (about 2% difference), while in York the use of two fonts was nearly two-fold that for one font. In Scarborough, the use of two fonts was also double that of the use of one font, but it contained more headstones with three fonts than with only one. This was very unlike the former two cities; the use of three fonts was not as popular with Oxford containing just over 20% and York just over 10%. The use of four fonts was very low for England, with Oxford containing the highest proportion at nearly 15%.

Table 1b: The number of different fonts found on individual headstones in the English churchyards examined

England	One Font	Two Fonts	Three Fonts	Four Fonts
Oxford	33.33%	31.31%	21.21%	14.14%
York	35.97%	51.80%	11.51%	1.00%
Scarborough	24.55%	41.32%	32.04%	2.69%

A χ^2 test was conducted to decipher whether there is a relationship with the number of font types on headstones in selected cities in the UK. The results show that there is no relationship between font type and city in the UK. However, examining the difference between the observed and expected numbers, there are two places where these are significantly different. In Scarborough, the observed number of three font types is over 20 counts higher than the expected, while in Edinburgh, for two font types, the expected is over 30 counts higher than the observed.

Apart from these, the observed and expected numbers that are of interest are between 10 and 20, and for York that is representative of one, two and, three font types. For Oxford these significant numbers rest in two and four font types; Dunbar shows a difference in two font types, and Scarborough and Edinburgh, besides the highest differences, have one-font type as a difference whereby the observed is higher than the expected.

DISCUSSION

The results found signify that York consistently contained a large enough difference (of around 10 counts minimum) between the observed and expected numbers for one to three font types. This could be attributed to the fact that two sites, St Denys's and Holy Trinity churchyards both had significantly lower numbers of headstones than were originally there. For the former, two-thirds of the original edifice were demolished in the 18th century (WYCCC, 2009). This suggests that any headstones that would have been present around those times would have been lost to these battles, or that the limited space available for burial played a role in the number of headstones that could be erected around this church. With time, this finite number would have lessened through weathering and vandalism. In this study, there was much evidence of headstones that were extremely weathered to the point that the only remaining evidence of these was their weathered stumps (Fig. **1**).

Figure 1: Photograph taken at St Mary Magdalen, Oxford, demonstrating a heavily weathered headstone (weathered stump).

One clear picture in this study is that there is a distinction between the number of four-font types employed. The results suggest that of the number of different fonts on a headstone, one and two are generally most commonly used; therefore, it is safe to say that the standard was one- or two-font types carved onto headstones in the UK between 1600 to the end of the Victorian period. According to Deetz and Dethlefsen (1971), the simplification of the motif was a product of the increased rate of production (p. 32); if this idea were applied to the simplification of headstone inscriptions with the use of one- or two-font types, then that would imply that there was more emphasis on mass production in Scotland than in England. With the more variety of styles in England having three- and four-font types, it could be assumed that there was more encouragement to produce headstones for the public rather than carve these to suit individual needs. Taking this basic idea and applying it to Scotland, it appears that there was more economic control, whereby there was far less variety in terms of the number of inscriptions used on any given headstone. This lack of originality was very common, and still is, in mass-produced products.

Moreover, in the same study (Deetz & Dethlefsen, 1971), it was mentioned that styles and variations on headstones, of motifs, were more common the further away from the main centre (p. 30). This could be drawn from what was fashionable at the time, as the elite were very much in control of the fashion trend, as it is evident today. In a sense this idea is evident in Oxford. Considering the city centre, Carfax, as a focal point, the churchyards with four font-types were furthest away from this point. Also, knowing that St Giles's churchyard was used as an overflow of burials from Oxford city (BHO, 2014), and in this churchyard although there were more headstones with two fonts, there was an equal proportion between one, three, and four. St Peter-in-the-East churchyard, located within the University of Oxford, contained no headstones with more than three font types, while all other churchyards contained three or more headstones with four font-types.

Aside from the number of inscriptions used, the structure of these is also important in navigating through a headstone for information and as a sign of standardisation during a period when these were being mass produced to avoid any unnecessary costs, particularly for those for whom these might have been too expensive so that specific words with a font type, such as Gothic font for introductions, were used. Eventually, this became the standard. According to Faraone and Rife (2007), the structure of inscriptions is crucial in the analysis of any material culture. Their examination of Greek lead curse tablets found in Kenchreai, Corinth demonstrated the use of specific verbs and nouns created in a sort of balance in the overall construction of these sorts of tablets (p. 154). In the same way, headstones have a structure that permit the reader to navigate the text in order to find important details, such as the name of the deceased or the place of birth or death; these were usually in upper-case or in Italics font. Moreover, on headstones, specific words or numbers were usually found in the same place on the façade of the headstone, as were most of the letterings on headstones post 1600 and early 1700s. In fact, headstones dated to the 1600s did not always follow the artful pattern used in subsequent years. Willsher (1985) noted that in Scotland in and prior to the early 1600s inscriptions were carved around the edges of the headstone to include all relevant details (p. 38).

In Mytum's (2004b) study, he found that a variety of upper- and lower-case lettering was apparent on headstones (p. 19); the former was true of the name of the deceased (or all those interred) and the latter of introductions and other carvings, in general. He also examined the differing inscriptions used in terms of size alongside the positioning of the different texts, such as dates and ages (p. 21).

According to his study, the dates and ages were either in Roman, or Italics fonts; whereas, in the latter, it was difficult to decipher some numbers. Mytum did not delve into the clarity of inscriptions here, particularly numbers, which are essential in this type of study.

According to Brears (1981), headstones in the Calder valley contained bold, simple inscriptions all in capitals and probably in Roman, and Serif fonts were used around the 1600s (p. 87). This is similar to this study in the sense that there was simplicity in the fonts used, where one type of font was employed, and a general simplicity in the use of two, with Gothic font dominating the introductions. Although these were not simple in any sense, their location and similarity made it simple to understand what they represented. According to Willsher (1985), Serif fonts were commonly used to artfully decorate the strokes at the top and bottom of lettering known as a flourish (p. 38), which was particularly noted in the introductions with 'Sacred' and 'In' from 'In memory of'. This became a commonality with non-Gothic lettering for the introductions. However, where Gothic lettering was used, there were no 'flourishes' noted. With the arrival of the use of the Gothic font, towards the late 18th century, words like 'Erected' or 'In memory of' were artfully carved at the top of headstones (Willsher, 1985, p. 38), to include 'Sacred' at least in this study.

The findings of this study conveyed that different fonts were used for different inscriptions on each headstone. Roman font was used for introductions, names of the deceased, and for entire inscriptions. The Roman font was also inscribed in varying sizes depending on the inscription, so that for a name this was often capitalised and slightly bigger than the surrounding font sizes. Gothic fonts were commonly used for introductions (Sacred to the memory of, Erected to/in the memory of, In memory of) and in some cases for linking inscriptions (also). Script font was commonly used for place names (location of death or birth) and for epitaphs. In some cases, however, entire inscriptions were in Script font. Fishtail font, to a lesser degree, was used in introductions (England) and for names (Scotland).

Gorman and DiBlasi (1981) believed that headstones, and possibly other monuments of commemoration, were a reflection of past attitudes towards death, family, and society (p. 79), and this is made quite evident in the varying styles of headstones found in burial grounds. Can the same be said about inscriptions? Are they as clearcut as motifs in the expression of ideals and values of past societies? This depends on how freely the headstone is used by the family of the deceased. According to Stewart (2007), inscriptions on headstones intimated the life of the dead, such as detailing the expression of feeling satisfaction in not having to face 'the stormy wind and tempest' that was the sea (p. 118). It is elements such as these that are important in understanding a people or community.

CONCLUSION

This study was meant to examine the number of fonts used on headstones throughout a 300-year period in selected cities within Scotland and England. The results conveyed that in Scotland the maximum fonts used on a given headstone were three, unlike England which had a maximum number of four. Moreover, common patterns of inscriptions for introductions, names, and dates were found, which not only encouraged the identification of important information off the headstone, but was allowed for the mass production of these monuments to meet the demands of the societies explored here.

ACKNOWLEDGEMENTS

In addition to the previous specific acknowledgements Chapter 7, the authors are grateful to the staff at all sites visited in this research.

CHAPTER 9

The Future

Abstract: Several changes made to digital photography have improved the use of the digital camera. However, there are still issues that affect digital photography in the field, and with continued innovation in this area for particular elements, such as camera shake, low battery, lighting conditions and obstructions, the use of the digital camera in the field could create a new trend in photographic research.

Keywords: Flash, lens, ghost-imaging, one-pixel camera, robotic tripod, vegetation cover.

The future of photography in research is bright; the age of technology supports the use of the digital photograph and programs used to examine the photograph, particularly to study details; researchers who study historical monuments or buildings could do so without having to trek through the field or observe an object in a protective space like in a gallery or museum. Despite using digital photography in the field, there are certain elements that hinder the capturing of a good photograph. Issues such as: 1) camera shake, 2) low battery, 3) lighting conditions, and 4) obstructions, have all, at one point or another, affected the quality of the picture being taken.

Camera Shake

Researchers at MIT have discovered a way to combat a shaking camera to sustain a usable image with the use of a robotic tripod to ensure that a clear image is produced (Barras, 2008, para. 9). This tool could take a panoramic view without any disruption in the movement of the camera. In headstone research, the issue that surfaces is tiredness; in many cases, the camera operator will be unable to hold the camera steady to ensure that the image being captured is what they need. In this case, a light-weight and portable, easy to assemble, tripod would be suitable with an easy automatic levelling device so that the time required to level the camera could be reduced by half.

Low Battery

In terms of battery life, this has been a significant factor in fieldwork. In several cases, fieldwork had to be stopped because the camera's battery could no longer sustain the energy requirements for taking a simple picture particularly when it is cold. Barras (2008) also wrote in *New Scientist* about the one pixel digital camera that is meant to improve battery performance (para. 23), especially when out in the field, where recharging facilities may not be available. However, the quality of the image may not be the best for research based on the sample image provided (Barras, 2008, para. 21). In this type of research, detail is extremely important. In some cases, researchers have brought their own charging devices, to include charging from their car. Having a solar power digital field camera would be suitable for the work involved, as long as the camera is light-weight and easy to use. Moreover, this would be suitable to fieldwork that is light-sensitive, such as photographing headstones.

Lighting Conditions

There are two conditions associated with lighting and these are extreme lighting conditions and, the opposite, poor light availability. The former is a rarity in the UK, but when on a sunny day a photograph is to be taken of a headstone that has resin or some sort of protective layer on it, this reflective surface can make taking a direct photograph challenging. In this case, a photograph at an angle would work best, but, unless required, these are not ideal (oblique). In this case, cameras that deal with glare without damaging the detail on the image would be ideal. Using software to remove this in a non-destructive way would be useful. Alternatively, according to Raskar at Mitsubishi Electric Research Laboratories in Cambridge, USA, as reported by Barras (2008), it is possible to eradicated glare through the use of a mask that fits on the camera's lens and image sensor (para. 14). This results in bright spots that could easily be eliminated through image processing.

In terms of poor lighting conditions, where it becomes difficult to decipher inscriptions on the headstone, software could be used to contrast these details on the photograph. Using a flash would resolve this, but the best conditions would be natural sunlight conditions, if possible as 'the flash can flatten out [the] digital image' and this compresses the depth of the image (Rowse, 2014, para. 4). Increasing the ISO (light sensitivity falling on the sensor) will mean that camera shake becomes an issue (Rowse, 2014, para. 6), which is something that has occurred in headstone research in the UK. Having additional light sources that act as natural light would probably be more useful than using the flash or a higher ISO. These light sources (or spot-lights) could be assembled around the object; but again, this would be too time-consuming and would not be very portable. Henceforth, more technology is needed to combat this issue, as it can be quite problematic in the field.

Obstructions

The final type of issue that has been experienced whilst photographing headstones has been vegetation covering, to include tree branches or trunks. Research had been conducted on the quantum camera by Shih from the University of Maryland, USA (Barras, 2008, para. 24); this camera could take ghost imaging of an object that might be obstructed by clouds or smoke. However, for this type of research, only if a camera could take an image of an object directly behind vegetation covering, would it be useful in the field. This would mean that several objects that had been overlooked could be researched and analysed.

A Final Point

These advances in technology, or the potential research that could be done on digital photography, could pave the way for researchers who have been focusing on preservation and conservation. Perhaps the age of improving the state of monuments *in situ* is slowly disappearing. Too many variables work against the sustainability of historical monuments that are exposed. With time, these too will vanish unless they are restored digitally, where they could be further studied by future scientists, who could then explore new methods of commemorating and honouring individuals. It is very evident that without digital photography, it is only time before headstones and other monuments of historical importance vanish permanently.

ACKNOWLEDGEMENTS

Declared None.

Chapter 10: CONCLUSION

CONCLUSION

Abstract: This book represents a collaborative work between a geomorphologist and archaeologist. The focus has been on the employment of photography to study and record the urban landscape. Both authors have brought their own disciplinary perspectives into the eBook, for pictorial depiction, image quantification, database compilation, and more. This collaboration is unique because of the theme (of photography), which is able to link the various research studies presented here. Photography has enabled science and social science to appear side-by-side in a harmonious contribution to landscape studies.

Keywords: Archaeology, geomorphology, fieldwork, interdisciplinary/cross-disciplinary research, landscape studies, urban environments.

Photographs are indispensable for capturing landscapes and their assessment. Traditionally, they have been conceived only qualitatively to portray pictorial depictions of landscapes; for instance, to show aspects of a study area. However, more recently, the photograph is being used for visualisation and as a recording devise to capture landscape change. In addition to air-based imagery, ground-based photography can be used to quantify landscape change through time. Close-up imagery is particularly useful to portray change, as for instance weathering patterns, vegetation cover, and so on in natural landscapes. This approach is developing in quantitative (re)photography at various spatial scales and in different landscapes, including urban environments.

This eBook has portrayed the use of photography in landscape studies, particularly focusing on cross-temporal change. Before-and-after comparisons are still popular, but quantitative photographic approaches are now added to this in capturing landscape change. The first-listed author has made various contributions to these areas within the urban setting of Oxford, UK, and various published works have been incorporated in this book. With the inclusion of a scale bar or colour chart, calibrated images can be quantitatively assessed and numerical results obtained for comparison with previous years of study or ongoing collected datasets.

Besides capturing pictorial depictions of landscapes, it is now possible to measure colour-based change, as of algae and lichens on limestone surfaces. This, in turn, can be deployed as evidence of microclimatic effects or change and

Mary J. Thornbush and Sylvia E. Thornbush

environmental quality. Photographs have also been useful in the assessment of vegetation change through time of various types of plant cover, including algae, lichens, and climbers. It is possible to employ photographic images at various scales and from different angles, as in planform to assess changes in vegetation (and perhaps land use) on the land surface. Imagery has been useful to present planetary characteristics of the land surface, as on Mars, and planetary applications are increasingly popular and no longer limited to the surface visualisation of planet Earth.

The use of photography has similarly been useful to the second-listed author. Its purposes are extremely important in recording historical monuments that are currently, and have been for some time, threatened to be destroyed through weathering, vandalism, and other forms of destruction. The need to preserve such structures is still prominent. Websites containing pictures of features, and other objects that have felt the effects of time, are accessible to the public; yet, these are examples of how much interest there is in these monuments.

In terms of preservation, the importance of containing such images in one specific area is needed for either genealogical, historical, or archaeological research. There is currently an interest in achieving this by the second-listed author. Ultimately, once a photographic database is established, it will be possible to study these monuments comfortably from anywhere in the world. Until other means of visual-record preservation (other than the video) could be established, photography will be at the foreground of aiding in the accessibility of such important cultural heritage.

ACKNOWLEDGEMENTS

Declared None.

REFERENCES

Aber, J. S., Aber, S. W., Buster, L., Jensen, W. E., & Sleezer, R. L. (2009). Challenge of infrared kite aerial photography: A digital update. *Transactions of the Kansas Academy of Science 112*(1, 2), 31-39.

Achterberg, R. A. (2007). *Photographs as primary sources for historical research and teaching in education: The Albert W. Achterberg Photographic Collection.* Retrieved 7 January 2014, from http://repositories.lib.utexas.edu/handle/2152/3538.

Agarwal, R. P. & Mitra, D. S. (1991). Paleogeographic reconstruction of Bengal Delta during Quaternary period. *Memoirs – Geological Society of India, 22*, 13-24.

Amoeda, R., Lira, S., & Pinheiro, C. (Eds.) (2012). Heritage 2012: 3rd International Conference on Heritage and Sustainable Development, 19-22 June 2012. Porto, Portugal: Green Lines Institute for Sustainable Development.

Andrew, C. (2002). Perception and aesthetics of weathered stone façades. In R. Přikryl & H. A. Viles (Eds.), *Understanding and managing stone decay* (pp. 331-339). Prague: The Karolinum Press.

Antill, S. J. & Viles, H. A. (1998). Deciphering the impacts of traffic on stone decay in Oxford: some preliminary observations from old limestone walls. In M. S. Jones & R. D. Wakefield (Eds.), *Stone Weathering and Atmospheric Pollution Network 1997: Aspects of stone weathering, decay and conservation* (pp. 28-42). Aberdeen.

Antrop, M. & Van Eetvelde, V. (2000). Holistic aspects of suburban landscapes: Visual image interpretation and landscape metrics. *Landscape and Urban Planning, 50*, 43-58.

Aprile, G. G., di Salvatore, M., Carratù, G., Mingo, A., & Carafa, A. M. (2010). Comparison of the suitability of two lichen species and one higher plant for monitoring airborne heavy metals. *Environmental Monitoring and Assessment, 162*, 291-299.

Arkell, W. J. (1947). *Oxford stone.* London: Faber & Faber.

Ashmolean Museum. (2012). *All Online Collections.* Retrieved 10 August 2014, from http://www.ashmolean.org/collections/online/.

Barras, C. (2008). *The Future of Photography.* NewScientist Tech, 16 September 2008. Retrieved from http://www.newscientist.com/article/dn14735-the-future-of-photography.html?full=true#.U-yjJPldVt0.

Bass, Jr, J. O. (2004). More trees in the tropics. *Area, 36*, 19-32.

Bergman, I., Östlund, L., Zackrisson, O., & Liedgren, L. (2007). Stones in the snow: A Norse fur traders' road into Sami country. *Antiquity, 81*, 397-408.

Bennett, L. T., Judd, T. S., & Adams, M. A. (2000). Close-range vertical photography for measuring cover changes in perennial grasslands. *Journal of Range Management, 53*, 634-641.

Bierman, P. R., Howe, J., Stanley-Mann, E., Peabody, M., Hilke, J., & Massey, C. A. (2005). Old images record landscape change through time. *GSA Today, 15*, 4-10.

Birch, G. P. (1990). Engineering geomorphological mapping for cliff stability. In Institute of Civil Engineers (ICE) & J. B. Burland (Eds.), *Chalk: Proceedings of the International Chalk Symposium* held at Brighton Polytechnic on 4-7 September 1989 (pp. 545-549). London: Thomas Telford for ICE.

Black Hart Entertainment. (2010). *City of the Dead Haunted Graveyard Tour.* Retrieved 24 May 2010, from http://www.cityofthedeadtours.com.

Bradwell, T. (2010). Studies on the growth of Rhizocarpon geographicum in NW Scotland, and some implications for lichenometry. *Geografiska Annaler, 92A*, 41-52.

Brears, P. C. D. (1981). Heart gravestones in the Calder Valley. *Folklife, 19*, 84-93.

British History Online (BHO). (2014). *Edinburgh – Edinburghshire.* Retrieved 13 August 2014, from http://www.british-history.ac.uk/report.aspx?compid=43437&strquery=1559%20AND%20preservation%20AND%20monument

Broadbent, N. D. (1990). Use of lichenometric and weathering rates for dating. *Norwegian Archaeological Review, 23*(1-2), 3-8.

Brook, B. W. & Bowman, D. M. J. S. (2006). Postcards from the past: Charting the landscape-scale conversion of tropical Australian savanna to closed forest during the 20th century. *Landscape Ecology, 21*, 1253-1266.

Brown, J. (2012). *The epitaphs and monumental inscriptions in Greyfriars Churchyard, Edinburgh.* Memphis: General Books LLC.

Brunialti G. & Frati, L. (2007). Biomonitoring of nine elements by the lichen *Xanthoria parietina* in Adriatic Italy: A retrospective study over a 7-year time span. *Science of the Total Environment, 387*, 289-300.

Buckham, S. (2002). New initiatives in the recording and the management of graveyards. In A. Dakin (Ed.), *Conservation of historic graveyards* (pp. 21-22). Edinburgh: Historic Scotland.

Burga, C. A., Frauenfelder, R., Ruffet, J., Hoelzle, M., & Kääb, A. (2004). Vegetation on alpine rock glacier surfaces: A contribution to abundance and dynamics on extreme plant habitats. *Flora, 199*, 505-515.

Burga, C. A., Krüsi, B., Egli, M., Wernli, M., Elsener, S., Ziefle, M., Fischer, T., & Mavris, C. (2010). Plant succession and soil development on the foreland of the Morteratsch glacier (Pontresina, Switzerland): Straight forward or chaotic? *Flora, 205*, 561-576.

Burgess, F. (1963). *English churchyard memorials.* London, UK: SPCK.

Burnside, N. G., Smith, R. F., & Waite, S. (2003). Recent historical land use change on the South Downs, United Kingdom. *Environmental Conservation, 30*, 52-60.

Butler, D. R. (1994). Repeat photography as a tool for emphasizing movement in physical geography. *Journal of Geography, 93*, 141-151.

Butler, D. R. & DeChano, L. M. (2001). Environmental change in Glacier National Park, Montana: An assessment through repeat photography from fire lookouts. *Physical Geography, 22*, 291-304.

Butler, D. R. & Malanson, G. P. (1990). Non-equilibrium geomorphic processes and patterns on avalanche paths in the northern Rocky Mountains, U.S.A. *Zeitschrift für Geomorphologie, 34*, 257-270.

Butler, D. R. & Malanson, G. P. (1993). An unusual early-winter flood and its varying geomorphic impact along a subalpine river in the Rocky Mountains of Montana, USA. *Zeitschrift für Geomorphologie, 37*, 145-155.

Camuffo, D. & Bernardi, A. (1993). Microclimatic factors affecting the Trajan Column. *Science of the Total Environment, 128*, 227-255.

Camuffo, D., Sturaro, G., & Valentino, A. (1999). Urban climatology applied to the deterioration of the Pisa Leaning Tower, Italy. *Theoretical and Applied Climatology, 63*, 223-231.

Cannon, A., Bartel, B., Bradley, R., Chapman, R. W., Curran, M. L., Gill, David W. J., Humphreys, S. C., Masset, Cl., Morris, I., Quilter, J., Rothschild, N. A., & Runnels, C. (1989). The historical dimension in mortuary expressions of status and sentiment [and comments and reply]. *Current Anthropology, 30*(4), 437-458.

Carignan, J. & Gariépy, C. (1995). Isotopic composition of epiphytic lichens as a tracer of the sources of atmospheric lead emissions in southern Québec, Canada. *Geochimica et Cosmochimica Acta, 59*(21), 4427-4433.

Carreras, H. A., Wannaz, E. D., Perez, C. A., & Pignata, M. L. (2005). The role of urban pollutants on the performance of heavy metal accumulation in *Usnea amblyoclada. Environmental Research, 97*, 50-57.

Carreras, H. A., Rodriguez, J. H., González, C. M., Wannaz, E. D., Garcia Ferreyra, F., Perrez, C. A., & Pignata, M. L. (2009). Assessment of the relationship between total suspended particles and the response to two biological indicators transplanted to an urban area in central Argentina. *Atmospheric Environment, 43*, 2944-2949.

de Castella, T. (2012). Five ways the digital camera changed us. *BBC News Magazine.* Retrieved 12 January 2012, from http://www.bbc.co.uk/news/magazine-16483509.

Chabot, N. J. (1996). A note about historical photography. In R. W. Preucel & I. Hodder (Eds.), *Contemporary archaeology in theory: A reader.* Oxford: Blackwell Publishers.

Chandler, J. (1999). Effective application of automated digital photogrammetry for geomorphological research. *Earth Surface Processes and Landforms, 24*, 51-63.

Clairmont, C. W. (1993). *Classical Attic tombstones.* Greece: Akanthus.

Clark, P. E. & Hardegree, S. P. (2005). Quantifying vegetation change by point sampling landscape photography time series. *Rangeland Ecology & Management, 58*, 588-597.

Clarke, J. P. (1965). Late eighteenth century decorated headstone at Orney Graveyard, Co. Louth. *The Journal of The County Louth Archaeological Society, 16*(1), 1-4.

Cooke, R. U., Inkpen, R. J., & Wiggs, G. F. S. (1995). Using gravestones to assess changing rates of weathering in the United Kingdom. *Earth Surface Processes and Landforms, 20*, 531-546.

Crang, M. & Cook, I. (2007). *Doing ethnographies.* London: SAGE.

Crimmins, M. A. & Crimmins, T. M. (2008). Monitoring plant phenology using digital repeat photography. *Environmental Management, 41*, 949-958.

Crossley, H. (1991). *Lettering in stone*. Upton-upon-severn: The Self Publishing Association.

Dahlberg, A. C. (2000). Landscape(s) in transition: An environmental history of a village in north-east Botswana. *Journal of Southern African Studies, 26*, 759-782.

Davidson, C. I., Tang, W., Finger, S., Etyemezian, V., Striegel, M. F., & Sherwood, S. I. (2000). Soiling patterns on a tall limestone building: Changes over 60 years. *Environmental Science and Technology, 34*, 560-565.

Dethlefsen, E. & Deetz, J. (1966) Death's heads, cherubs, and willow trees: Experimental archaeology in colonial cemeteries. *American Antiquity, 31*(4), 502-510.

Deetz, J. & Dethlefsen, E. N. (1971). Some social aspects of New England colonial mortuary art. *Memoirs of the Society for American Archaeology, 25*, 30-38.

Dragovich, D. (1987). Measuring stone weathering in cities: Surface reduction on marble monuments. *Environmental Geology and Water Sciences, 9*, 139-141.

Elliott, G. P. & Baker, W. L. (2004). Quaking aspen (*Populus tremuloides* Michx.) at treeline: A century of change in the San Juan Mountains, Colorado, USA. *Journal of Biogeography, 31*, 733-745.

English Heritage. (2011). *ViewFinder*. Retrieved 31 March 2007, from http://viewfinder.english-heritage.org.uk/.

Erol, O. (1999). A geomorphological study of the Sultansazligi lake, central Anatolia. *Quaternary Science Reviews, 18*(4-5), 647-657.

Estrabon, C., Filippini, E., Soria, J. P., Schelotto, G., & Rodriguez, J. M. (2011). Air quality monitoring system using lichens as bioindicators in Central Argentina. *Environmental Monitoring and Assessment, 182*, 375-383.

Evans, D. J. A., Archer, S. & Wilson, D. J. H. (1999). A comparison of the lichenometric and Schmidt hammer dating techniques based on data from the proglacial areas of some Icelandic glaciers. *Quaternary Science Reviews, 18*, 13-41.

Faraone, C. A. & Rife, J. L. (2007). A Greek curse against a thief from the Koutsongila Cemetery at Roman Kenchreai. *Zeitschrift für Papyrologie und Epigraphik*, 141-157.

Fassina, V., Favaro, M., & Naccari, A. (2002). Principal decay patterns on Venetian monuments. In S. Siegesmund, T. Weiss & A. Vollbrecht (Eds.), *Natural stone, weathering phenomena, conservation strategies and case studies* (Special Publications 205, pp. 381-391). London: Geological Society.

Ferguson, M. (2002). Record-taking and record-keeping for graveyards. In A. Dakin (Ed.), *Conservation of historic graveyards* (pp. 8-9). Edinburgh: Historic Scotland.

Fitzpatrick, F. (1980). Inscriptions in Galoon Cemetery. *Clogher Record, 10*(2), 264-268.

Fletcher, R. & St George, J. (n.d.). *Strengths of Young Parents Project: Rationale for using images to talk to young parents about their children*. Newcastle: Interrelate Family Centres, University of Newcastle.

Foote, K. E. (1985). Velocities of change of a built environment, 1880–1980: Evidence from the photoarchives of Austin, Texas. *Urban Geography, 6*, 220-245.

Foster, S. (2002). Carved stones policy – New developments. In A. Dakin (Ed.), *Conservation of historic graveyards* (pp. 1-2). Edinburgh: Historic Scotland.

Foster, N. H. & Beaumont, E. A. (1992). Photogeology and photogeomorphology. In N. H. Foster & E. A. Beaumont (Eds.), *Photogeology and photogeomorphology*. London: American Association of Petroleum Geologists.

Fraser, A. (2002). Integrated management: Mount Auburn Cemetery. In A. Dakin (Ed.), *Conservation of historic graveyards* (pp. 17-18). Edinburgh: Historic Scotland.

Fraser P. M. & Rönne, T. (1957). *Boeotian and West Greek tombstones*. Sweden: C. W. K. Gleerup.

Frith, H. & Harcourt, D. (2007). Using photographs to capture women's experiences of chemotherapy: Reflecting on the method. *Qualitative Health Research, 17*(10), 1340-1350.

Garty, J., Levin, T., Cohen, Y., & Lehr, H. (2002). Biomonitoring air pollution with the desert lichen *Ramalina maciformis*. *Physiologia Plantarum, 115*, 267-275.

Geological Survey of India (1982). *Proceedings of the workshop on problems of the deserts in India*. Jaipur.

Gittings, C. (2002). Bereavement and commemoration: An archaeology of mortality by Sarah Tarlow. *Folklore, 113*(1), 120-121.

Gob, F., Petit, F., Bravard, J.- P., Ozer, A., & Gob, A. (2003). Lichenometric application to historical and subrecent dynamics and sediment transport of a Corsican stream (Figarella River—France). *Quaternary Science Reviews, 22*, 2111-2124.

Gorman, F. J. E. & DiBlasi, M. (1981). Gravestone iconography and mortuary ideology. *Ethnohistory, 28*, 79-98.

Goudie, A. S. (1986). Laboratory simulation of 'the wick effect' in salt weathering of rock. *Earth Surface Processes and Landforms, 11*(3), 275-285.

Griffin, R. D., Stahle, D. W., & Therrell, M. D. (2005). Repeat photography in the ancient Cross Timbers of Oklahoma, USA. *Natural Areas Journal, 25*, 176-182.

Griffith Institute. (2005). *Photographs by Harry Burton*. Retrieved 10 August 2014, from http://www.griffith.ox.ac.uk/gri/carter/gallery/.

Hall, D. W. (2001). Signs of the dead: Epitaphs, inscriptions, and the discourse of the self. *ELH, 68*(3), 655-677.

Hall, F. C. (2001). *Ground-based photographic monitoring*. General Technical Reports of the US Department of Agriculture, Forest Service, pp. 1-75.

Hall, F. (2007). Photomonitoring. Proceedings: International Conference on Transfer of Forest Science Knowledge and Technology, General Technical Report PNW, 726, pp. 93-101.

Hamann, S., Herman, R. A., Nolan, C. L., & Wallen, K. (2004). Men and women differ in amygdala response to visual sexual stimuli. *Nature Neuroscience*, 7, pp. 411-416. Retrieved from http://www.nature.com/neuro/journal/v7/n4/abs/nn1208.html.

Hansen, E. S., Dawes, P. R., & Thomassen, B. (2006). Epilithic lichen communities in High Arctic Greenland: Physical, environmental, and geological aspects of their ecology in Inglefiled Land (78°–79°N). *Arctic, Antarctic, and Alpine Research, 38*(1), 72-81.

Harper, D. (2002). Talking about pictures: A case for photo elicitation. *Visual Studies, 17*(1), 13-26.

Hendren, J. W. (1938). Epitaphs from down east. *The New England Quarterly, 11*(3), 524-540.

Hereford, R., Webb, R. H., & Longpré, C. I. (2006). Precipitation history and ecosystem response to multidecal precipitation variability in the Mojave Desert region, 1893–2001. *Journal of Arid Environments, 67*, 13-34.

Hijiya, J. A. (1983). American gravestones and attitudes toward death: A brief history. *Proceedings of the American Philosophical Society, 127*(5), 339-363.

Historic Scotland. (n.d.). *Why is the Historic Environment Important?*. Retrieved from http://www.historic-scotland.gov.uk/index/heritage/valuingourheritage/why-is-the-historic-environment-important.htm.

Hoke, G. D. & Turcotte, D. L. (2004). The weathering of stones due to dissolution. *Environmental Geology, 46*, 305-310.

Hope, V. M. (1997). Words and pictures: The interpretation of Romano-British tombstones. *Britannia, 28*, 245-258.

Hudak, A. T. & Wessman, C. A. (1998). Textural analysis of historical aerial photography to characterize woody plant encroachment in South Africa savanna. *Remote Sensing of Environment, 66*, 317-330.

Inkpen, R. J. & Jackson, J. (2000). Contrasting weathering rates in coastal, urban and rural areas in southern Britain: Preliminary investigations using gravestones. *Earth Surface Processes and Landforms, 25*, 229-238.

Inkpen R. J., Fontana D., & Collier P. (2001). Mapping decay: integrating scales of weathering within a GIS. *Earth Surface Processes and Landforms, 26*(8), 885-900.

Inkpen R., Duane B., Burdett J., & Yates T. (2008). Assessing stone degradation using an integrated database and geographical information system (GIS). *Environmental Geology, 56*(3-4), 789-801.

Innes, J. L. (1985). An examination of some factors affecting the largest lichens on a substrate. *Arctic and Alpine Research, 17*(1), 99-106.

Isocrono, D., Matteucci, E., Ferrarese, A., Pensi, E., & Piervittori, R. (2007). Lichen colonization in the city of Turin (N Italy) based on current and historical data. *Environmental Pollution, 145*, 258-265.

Jacoby Petersen, N. & Ostergaard, S. (2003). Organisational photography as a research method: What, how and why. Paper presented in The Academy of Management Conference Proceedings, Seattle.

Jenkins, S., (n.d.). *Burial Grounds: Churchyards and Cemeteries*. Retrieved 10 July 2014, from http://oxfordhistory.org.uk/burials/burial_grounds/index.html.

Jomelli, V. Grancher, D., Naveau, P., Cooley, D., & Brunstein, D. (2007). Assessment study of lichenometric methods for dating surfaces. *Geomorphology, 86*, 131-143.

Jones, J. (2009). *St Cross Church, Holywell: Burials and Inscriptions*. Retrieved 24 May 2010, from http://archives.balliol.ox.ac.uk/Archives/archivesmss.asp.

Jones, S. (2000). 'Enlivening' development concepts through workshops: A case study of appropriate technology and soil conservation. *Journal of Geography in Higher Education, 24*(1), 76-86.

Jovan, S. & McCune, B. (2005). Air-quality bioindication in the greater Central Valley of California, with epiphytic macrolichen communities. *Ecological Applications, 15*(5), 1712-1726.

Kadmon, R. & Harari-Kremer, R. (1999). Studying long-term vegetation dynamics using digital processing of historical aerial photographs. *Remote Sensing of Environment, 68*, 164-176.

Kantvilas, G., Elix, J. A., & James, P. W. (1992). Siphulella, a new lichen genus from Southwest Tasmania. *Bryologist, 95*(2), 186-191.

Keister, D. (2004). *Stories in stone: A field guide to cemetery symbolism and iconography*. Layton: Gibbs Smith.

Kellerhals, R., Church, M., & Bray, D. I. (1976). Classification and analysis of river processes. *Journal of the Hydraulics Division, 102*, 813-829.

Kelly, T. E. (1961). Photogeology – Quick, economical tool for oil hunters. *Oil and Gas Journal, 59*(47), 265-266.

Kidron, G. J. & Temina, M. (2010). Lichen colonisation on cobbles in the Negev Desert following 15 years in the field. *Geomicrobiology Journal, 27*, 455-463.

Kightly, C. (2004). *Church of Saint Denys, Walmgate, York: Historical notes and visitors guide*. York: St Denys Church.

Kselman, T. A. (2014). *Death and afterlife in modern France*. Princeton: Princeton University Press.

Kull, C. A. (2005). Historical landscape repeat photography as a tool for land use change research. *Norwegian Journal of Geography, 59*, 253-268.

Kullman, L. (2006). Long-term geobotanical observations of climate change impacts in the Scandes of West-Central Sweden. *Nordic Journal of Botany, 24*, 445-467.

Kutiel, P., Lavee, H., & Ackermann, O. (1998). Spatial distribution of soil surface coverage on north and south facing hillslopes along a Mediterranean to extreme arid climatic gradient', *Geomorphology, 23*, 245-256.

Lammel, G. & Metzig, G. (1997). Pollutant fluxes onto the façades of a historical monument. *Atmospheric Environment, 31*(15), 2249-2259.

Lane, S. N., Richards, K. S., & Chandler, J. H. (1993). Developments in photogrammetry: The geomorphological potential. *Progress in Physical Geography, 17*, 306-328.

Langford, M. (1978). *The step by step guide to photography: A complete manual*. London: Ebury Press.

Larsen, R. S., Bell, J. N. B., James, P. W., Chimonides, P. J., Rumsey, F. J., Tremper, A., & Purvis, O. W. (2007). Lichen and bryophyte distribution on oak in London in relation to air pollution and bark acidity. *Environmental Pollution, 146*, 332-340.

Liamputtong, P. (2007). *Researching the vulnerable: A guide to sensitive research methods*. London: SAGE.

Lind, A. O. (1974). *Photo-geomorphology of coastal landforms, Cat Island, Bahamas*. Belvoir: Army Engineer Topographic Labs.

Linz, S. (2014). An overview of photography as a creative modality in qualitative research. PowerPoint presentation used in a Seton Hall University project for graduate students of nursing, South Orange.

Longfield, A. K. (1946). Some 18th century Irish tomb-stones (continued). *Journal of the Royal Society of Antiquaries of Ireland, 76*, 81-88.

Longfield, A. K. (1954). Some 18th century Irish tombstones (continued). *Journal of the Royal Society of Antiquaries of Ireland, 84*, 173-178.

Longfield, A. K. (1955). 18th century Irish tombstones (continued). *Journal of the Royal Society of Antiquaries of Ireland, 85*, 114-117.

Ludwig, A. (1966). *Graven images: New England stonecarving and its symbols, 1650-1815*. Hanover: University Press of New England.

Manier, D. J. & Laven, R. D. (2002). Changes in landscape patterns associated with the persistence of aspen (*Populus tremuloides* Michx.) on the western slope of the Rocky Mountains, Colorado. *Forest Ecology and Management, 167*, 263-284.

McCabe, S., Smith, B. J., & Warke, P. A. (2007). Preliminary observations on the impact of complex stress histories on sandstone response to salt weathering: Laboratory simulations of process combinations. *Environmental Geology, 52*, 269-276.

McCarthy, M. (2009, October 7). Scottish graveyards among world's most at-risk historic sites, *The Independent* (News), p. 14.

McKendrick, J. H. & Bowden, A. (1999). Something for everyone? An evaluation of the use of audio-visual resources in geographical learning in the UK. *Journal of Geography in Higher Education, 23*(1), 9-19.

McKendrick, J. H. & Bowden, A. (2000). Equipped for the 21st century?: Audio-visual resource standards and product demands from geography departments in the UK. *Journal of Geography in Higher Education, 24*(1), 53-73.

McKendry, J. (2003). *Into the silent land: Historic cemeteries and graveyards in Ontario.* Kingston: Jennifer McKendry.

McMahon, R. (2002). The North Burial Ground Gravestone Restoration Project. In A. Dakin (Ed.), *Conservation of historic graveyards* (p. 16). Edinburgh: Historic Scotland.

McNeill, G. W. (1999). Variations in the weathering rate of Scottish gravestones as an environmental signature of atmospheric pollution. *Environmental Geochemistry and Health, 21,* 365-370.

Meyer, E. A. (1990). Explaining the epigraphic habit in the roman empire: the evidence of epitaphs. *The Journal of Roman Studies, 80,* 74-96.

Miller, V. C. (1968). Aerial photographs and surface features – 1. Aerial photographs and land forms (photogeomorphology). In Centre national de la recherche scientifique (France) & Université de Toulouse (France), *Aerial surveys and integrated studies: Proceedings of the Toulouse Conference* held in Proceedings of the Toulouse Conference, 21-28 September 1964 (pp. 41-69). Paris: UNESCO.

Mitchell, R. & Gu, J. (2000). Changes in the biofilm microflora of limestone caused by atmospheric pollutants. *International Biodeterioration and Biodegradation, 46,* 299-203.

Mollard, J. D. (1973). *Landforms and surface materials of Canada: A stereoscopic airphoto altas and glossary* (4th ed.). Regina: Airphoto Interpretation, Regina.

Mollard, J. D. & Janes, J. R. (1984). *Airphoto interpretation and the Canadian landscape.* Hull: Canadian Government Publishing Centre.

Monte, M. (1991). Multivariate analysis applied to the conservation of monuments: Lichens on the Roman Aqueduct Anio Vetus in S. Gregorio. *International Biodeterioration, 28,* 133-150.

Mottershead, D. N., Bailey, B., Collier, P., & Inkpen, R. J. (2003). Identification and quantification of weathering by plant roots. *Building and Environment, 38,* 1235-1241.

Mould D. R. & Loewe, M. (2006). *Historic gravestone art of Charleston, South Carolina, 1695–1802.* Jefferson, SC: McFarland & Co.

Munroe, J. S. (2003). Estimates of Little Ice Age climate interred through historical photography, northern Uinta Mountains, U.S.A. *Arctic, Antarctic, and Alpine Research, 35,* 489-498.

Mustoe, G. E. (1982). The origin of honeycomb weathering. *Geological Society of America Bulletin, 93,* 108-115.

Mytum, H. (1994). Language as symbol in churchyard monuments: the use of welsh in nineteenth- and twentieth-century pembrokeshire. *World Archaeology, 26*(2), 252-267.

Mytum, H. (2000). *Recording and analysing graveyards.* York: Council for British Archaeology.

Mytum, H. (2004a). Artefact biography as an approach to material culture: Irish gravestones as a material form of genealogy. *The Journal of Irish Archaeology, 12-13,* 111-127.

Mytum, H. (2004b). Local traditions in early eighteenth-century commemoration: The headstone memorials from Balrothery, Co. Dublin, and their place in the evolution of Irish and British commemorative practice. *Proceedings of the Royal Irish Academy, 104C*(1), 1-35.

Mytum H. & Evans, R. (2002). The evolution of an Irish graveyard during the 18th century: The example of Killeevan, Co. Monaghan. *The Journal of Irish Archaeology, 11,* 131-146.

Nassar, I. (2006). Familial snapshots: representing Palestine in the work of the first local photographers. *History and Memory,* 18(2), 139-155.

Niyogi, D. (1988). Photogeomorphic and photopedologic mapping of Quaternary formations around Kalna Town, Dist. Barddhaman, W. Bengal. *Indian Journal of Earth Sciences, 15*(3), 216-227.

Nugari, M. P., Pietrini, A. M., Caneva, G., Imperi, F., & Visca, P. (2009). Biodeterioration of mural paintings in a rocky habitat: The Crypt of the Original Sin (Matera, Italy). *International Biodeterioration and Biodegradation, 63,* 705-711.

Nüsser, M. (2001). Understanding cultural landscape transformation: A re-photographic survey in Chitral, eastern Hindukush, Pakistan. *Landscape and Urban Planning, 57,* 241-255.

Oakeshott, W. F. (1975). *Oxford stone restored: The work of the Oxford Historic Buildings Fund, 1957–1974.* Oxford: Oxford University Press.

O'Daly, B. (1954). Tydavnet Old Cemetery. *Clogher Record, 1*(2), 43-55.

Ode, A., Tveit, M. S., & Fry, G. (2010). Advantages of using different data sources in assessment of landscape change and its effect on visual scale. *Ecological Indicators, 10*, 24-31.

Oxford City Council. (2014). *Oxford Airwatch Data Archive.* Retrieved 23 June 2011, from http://www.oxford-airwatch.aeat.co.uk/archive.php.

Paine, C. (1992). Landscape management of abandoned cemeteries in Ontario. *APT Bulletin. Conserving Historic Landscapes, 24*, 59-68.

Park, S. (2012, July 6). Lehi child dies after headstone falls on top of him. *Deseret News.* Retrieved from http://www.deseretnews.com/article/865558601/Lehi-child-dies-after-headstone-falls-on-top-of-him.html?pg=all.

Pech, D., Condal, A. F., Bourget, E., & Ardisson, P.-L. (2004). Abundance estimation of rocky shore invertebrates at small spatial scale by high-resolution digital photography and digital image analysis. *Journal of Experimental Marine Biology and Ecology, 299*, 185-199.

Petre, J. (2006, May 31). Old graves may be used for double and 'vertical' burials, *The Daily Telegraph* (News).

Pictures in colour of Oxford (c. 1907). Norwich: Jarrold & Sons.

Pink, S. (2012). Advances in visual methodology: An introduction. In S. Pink (Ed.), *Advances in visual methodology*. London: SAGE.

Pink, S. (2013). *Doing visual ethnography*. London; SAGE.

Poyntzpass and District Local History Society (PDLHS). (2000). Appendix 1: Acton (Old) Graveyard: Headstone inscriptions. *"Before I forget": Journal of the Poyntzpass and District Local History Society*, 90-100.

Proulx, R. & Parrott, L. (2009). Structural complexity in digital images as an ecological indicator for monitoring forest dynamics across scale, space and time. *Ecological Indicators, 9*, 1248-1256.

Ralston Art Memorial (n.d.). *What is the process done on carving the letters and pictures on to a headstone?.* Retrieved 12 August 2014, from http://ralstonartmemorial.com/process-headsone-carving/.

Rango, A., Huenneke, L., Buonopane, M., Herrick, J. E., & Havstad, K. M. (2005). Using historic data to assess effectiveness of shrub removal in southern New Mexico. *Journal of Arid Environments, 62*, 75-91.

Ray, J. L. & Smith, A. D. (2012). Using photographs to research organizations: Evidence, considerations, and application in a field study. *Organizational Research Methods, 15*(2), 288-315.

Reddy, M. M. (1988). Acid rain damage to carbonate stone: a quantitative assessment based on the aqueous geochemistry of rainfall runoff from stone. *Earth Surface Processes and Landforms, 13*, 335-354.

Rivard, L. A. (2011). *Satellite geology and photogeomorphology: An instructional manual for data integration.* Saint-Lambert: Springer.

Robb, J. E. (1998). The archaeology of symbols. *Annual Review of Anthropology, 27*, 329-346.

Rodwell, W. (1981). *The archaeology of the English church: Study of historic churches and churchyards.* London: Batsford.

Rose, G. (2000). Practising photography: An archive, a study, some photographs and a researcher. *Journal of Historical Geography, 26*, 555-571.

Rose, G. (2008). Using photographs as illustrations in human geography. *Journal of Geography in Higher Education, 32*, 151-160.

Rose, G. (2011). Making photographs as part of a research project: Photo-documentation, photo-elicitation and photo-essays. *Visual methodologies: An introduction to researching with visual materials* (3rd ed., pp. 297-327). Milton Keynes: SAGE.

Rossbach, M., Jayasekera, R., Kniewald, G., & Thang, N. H. (1999). Large scale air monitoring: Lichen *vs.* air particulate matter analysis. *The Science of the Total Environment, 232*, 59-66.

Rowse, D. (2014). *How to Get Better Digital Photos in Low Light Conditions Without Using a Flash* [Web log post]. Retrieved from http://digital-photography-school.com/how-to-get-better-digital-photos-in-low-light-conditions-without-using-a-flash/.

Ryan, J. R. (1997). *Picturing empire: Photography and the visualization of the British Empire.* London: Reaktion Books.

Sanchez-Eppler, K. (1988). Decomposing: Wordsworth's poetry of epitaph and English burial reform. *Nineteenth-Century Literature, 42*(4), 415-431.

Schwartz, D. (1989). Visual ethnography: Using photography in qualitative research. *Qualitative Sociology, 12*(2), 119-154.

Searle, D. E. (2001). *The comparative effects of diesel and coal particulate matter on the deterioration of Hollington sandstone and Portland limestone.* Unpublished doctoral thesis, University of Wolverhampton, UK.

Shukla, V. & Upreti, D. K. (2011). Changing lichen diversity in and around urban settlements of Garhwal Himalayas due to increasing anthropogenic activities. *Environmental Monitoring and Assessment, 174*, 439-444.

Sidaway, J. D. (2002). Photography as geographical fieldwork. *Journal of Geography in Higher Education, 26*(1), 95-103.

Smith, B. J. (1996). Scale problems in the interpretation of urban stone decay. In B. J. Smith & P. A. Warke (Eds.), *Processes of urban stone decay*, (pp. 3-18). Shaftesbury: Donhead Publishing.

Smith, L., Shackel, P., & Campbell, G. (Eds.) (2011). *Heritage, labour and the working class.* Abingdon: Routledge.

Smith, R. M., Thompson, K., Warren, P. H. & Gaston, K. J. (2010). Urban domestic gardens (XIII): Composition of the bryophyte and lichen floras, and determinants of species richness. *Biological Conservation, 143*, 873-882.

Snell, K. D. M. (2003). Gravestones, belonging and local attachment in England 1700–2000. *Past & Present,* 97-134.

Sontag, S. (1977). *On photography.* London: Penguin.

Stewart, D. J. (2007). Gravestones and monuments in the maritime cultural landscape: Research potential and preliminary interpretations. *The International Journal of Nautical Archaeology, 36*(1), 112-124.

Strangstad, L. (1988). *A graveyard preservation primer.* Walnut Creek, CA: AltaMira.

Swallow, P., Dallas, R., Jackson, S., & Watt, D. (2004). *Measurement and recording of historic buildings.* Shaftesbury: Donhead Publishing.

Swetnam, T. W., Allen, C. D., & Betancourt, J. L. (1999). Applied historical ecology: Using the past to manage for the future. *Ecological Applications, 9*, 1189-1206.

Talukdar, S. N. (1980). Application of remote sensing techniques to petroleum exploration in India. *Proceedings Workshop 22nd Plenary Meeting of COSPAR*, 29 May to 9 June 1979, pp. 121-126. Bangalore, India: Pergamon Press.

Tarlow, S. (1999). *Bereavement and commemoration: An archaeology of mortality.* Oxford: Blackwell Publishers.

Tator B. A. *et al.* (1960). Photo interpretation in geology. In R. N. Colwell (Ed.), *Manual of photographic interpretation* (pp 169-342). Falls Church: American Society of Photogrammetry.

Thornbury, W. D. (1970). *Principles of geomorphology* (2nd ed.). London: John Wiley and Sons.

Thornbush, M. (2008a). Postcards used to track environmental history. *Environmental History, 13*(2), 360-365.

Thornbush, M. (2008b). Grayscale calibration of outdoor photographic surveys of historical stone walls in Oxford, England. *Color Research and Application, 33*(1), 61-67.

Thornbush, M. J. (2010a). Measurements of soiling and colour change using outdoor rephotography and image processing in Adobe Photoshop along the southern façade of the Ashmolean Museum, Oxford. In B. J. Smith, M. Gomez-Heras, H. A. Viles, & J. Cassar (Eds.), *Limestone in the built environment: Present-day challenges for the preservation of the past* (Special Publications 331, pp. 231-236). London: Geological Society.

Thornbush, M. J. (2010b). Photographic surveys of building exteriors in central Oxford, UK. *International Journal of Architectural Heritage, 4*(4), 351-369.

Thornbush, M. J. (2012). Developing a weathering scale for limestone walls in central Oxford, UK. *Geosciences, 2*, 277-297.

Thornbush, M. J. (2013a). A photo-based environmental history of the use of climbing plants in central Oxford, UK. *International Journal of Geosciences, 4*(7), 1083-1094.

Thornbush, M. J. (2013b). Digital photography used to quantify the greening of north-facing walls along Broad Street in central Oxford, UK/ L'utilisation de la photographie numérique pour quantifier le verdissement de la façade septentrionale longeant Broad Street dans le centre d'Oxford, Royaume-Uni. *Géomorphologie: Relief, Processus, Environnement, 2*, 111-118.

Thornbush, M. J. (2013c). Tracking the use of climbing plants in the urban landscape through the photoarchives of two Oxford colleges, 1861–1964. *Landscape Research, 38*(3), 312-328.

Thornbush, M. J. (2013d). Photogeomorphological studies of Oxford stone – A review. *Landform Analysis, 22,* 111-116.

Thornbush, M. J. (2014). Evidence of climate change from an urban environment comprising long-term climatic records. In C. B. Keyes & and O. C. Lucero (Eds.), *New developments in global warming research* (pp. 133-142). Hauppauge, NY: Nova Science.

Thornbush, M. J. (2014a). Measuring surface roughness through the use of digital photography and image processing. *International Journal of Geosciences, 5*(5), 540-554.

Thornbush, M. J. (2014b). A soiling index based on quantitative photography at Balliol College in central Oxford, UK. *Journal of Earth, Ocean and Atmospheric Sciences, 1*(1), 1-15.

Thornbush, M. J. & Thornbush, S. E. (2013). The application of a limestone weathering index at churchyards in central Oxford, UK. *Applied Geography, 42,* 157-164.

Thornbush, M. & Viles, H. (2004a). Integrated digital photography and image processing for the quantification of colouration on soiled surfaces in Oxford, England. *Journal of Cultural Heritage, 5*(3), 285-290.

Thornbush, M. J. & Viles, H. A. (2004b). Surface soiling pattern detected by integrated digital photography and image processing of exposed limestone in Oxford, England. In C. Saiz-Jimenez (Ed.), *Air pollution and cultural heritage* (pp. 221-224). London: A. A. Balkema Publishers.

Thornbush, M. & Viles, H. (2005). The changing façade of Magdalen College, Oxford: Reconstructing long-term soiling patterns from archival photographs and traffic records. *Journal of Architectural Conservation, 11*(2), 40-57.

Thornbush, M. & Viles, H. (2006). Changing patterns of soiling and microbial growth on building stone in Oxford, England after implementation of a major traffic scheme. *Science of the Total Environment, 367*(1), 203-211.

Thornbush, M. J. & Viles, H. A. (2007a). Photo-based decay mapping of replaced stone blocks on the boundary wall of Worcester College, Oxford. In R. Přikryl & B. J. Smith (Eds.), *Building stone decay: From diagnosis to conservation* (Special Publications 271, pp. 69-75). London: Geological Society.

Thornbush, M. J. & Viles, H. A. (2007b). Simulation of the dissolution of weathered versus unweathered limestone in carbonic acid solutions of varying strength. *Earth Surface Processes and Landforms, 32*(6), 841-852.

Thornbush, M. J. & Viles, H. A. (2008). Photographic monitoring of soiling and decay of roadside walls in Oxford, England. *Environmental Geology, 56*(3-4), 777-787.

Tinkler, P. (2013). *Using photographs in social and historical research.* London: SAGE.

Tømmervik, H., Johansen, M. E., Pedersen, J. P., & Guneriussen, T. (1998). Integration of remote sensed and in-situ data in an analysis of the air pollution effects on terrestrial ecosystems in the border areas between Norway and Russia. *Environmental Monitoring and Assessment, 49,* 51-85.

The University Church of St Mary the Virgin. (2009). *The University Church of St Mary the Virgin.* Retrieved 23 June 2011, from http://www.university-church.ox.ac.uk/.

Urquhart, D. C. M. (2002). Brief description of the new Historic Scotland Guide for Practitioners – The conservation of historic graveyards. In A. Dakin (Ed.), *Conservation of historic graveyards* (p. 6). Edinburgh: Historic Scotland.

Victoria & Albert Museum (V&A). (2014a). *Search the Collections.* Retrieved 10 August 2014, from http://collections.vam.ac.uk/.

Victoria & Albert Museum (V&A). (2014b). *A History of the Portrait Miniature.* Retrieved from http://www.vam.ac.uk/content/articles/h/a-history-of-the-portrait-miniature/.

Viles, H. A. (1993). The environmental sensitivity of blistering of limestone walls in Oxford, England: A preliminary study. In D. S. G. Thomas & R. J. Allison (Eds.), *Landscape sensitivity* (pp. 308-326). Chichester: John Wiley and Sons Ltd.

Viles, H. A. (1994). Time and grime: Studies in the history of building stone decay in Oxford. Research Paper No. 50. Oxford: School of Geography, University of Oxford.

Viles, H. (1996). "Unswept stone, besmear'd by sluttish time": Air pollution and building stone decay in Oxford, 1790–1960. *Environment and History, 2*(3), 359-372.

Viles, H. A. (2001). Scale issues in weathering studies. *Geomorphology, 41,* 63-72.

Voutsaki, S. (2001). Bereavement and commemoration: An archaeology of mortality by Sarah Tarlow. *Journal of Archaeology, 105*(1), 112-113.

Vovelle, M. (1980). A century and one-half of american epitaphs (1660-1813): Toward the study of collective attitudes about death. *Comparative Studies in Society and History, 22*(4), 534-547.

Walker, L. (1983). Review: London cemeteries: An illustrated guide and gazetteer. *Journal of the American Society of Architectural Historians, 42*, 393-394.

Wang, Q., Lim, K. C., & Woo, H. L. (2006). Exploring the use of color photographs in Chinese picture composition writings: An action research in Singapore schools. *New Horizons in Education*, 91-102.

Warke, P. A. (1996). Inheritance effects in building stone decay. In B. J. Smith & P. A. Warke (Eds.), *Processes of urban stone decay* (pp. 32-43). London: Donhead Publishing.

Webb, R. H. (1996). *Grand Canyon, a century of change: Rephotography of the 1889–1890 Stanton Expedition.* Tuscon: The University of Arizona Press.

Webb, R. H., Boyer, D. E., & Turner, R. M. (Eds.). (2010). *Repeat photography: Methods and applications in the natural sciences.* Washington, DC: Island Press.

Welcome to St Mary's Church, Scarborough (WSMCS). (2007). Scarborough: St Mary's Church.

Welcome to York City Centre Churches (WYCCC). (2009). York: CoRE.

Willsher, B. (1985). *Understanding Scottish gravestones.* Edinburgh: Canongate Books.

Woodward Family Tree. (2013, September 30). A Guide to Church Burials. Retrieved on 20 August 2014 from http://www.gwoodward.co.uk/guides/burials.htm.

Wright, G. N. (1996). *Discovering epitaphs.* Princes Risborough: Shire Publications.

Young, A. R. M. (1987). Salt as an agent in the development of cavernous weathering. *Geology, 15*, 962-966.

Subject Index

A

Adobe photoshop 33, 64, 69
Aerial photographs 6, 8, 13
   encompassed 5
   including 35
   used 13
   using 13
Aerial photography 6, 8
   emphasised 6
   infrared kite 6
Air pollution 15, 28, 64-66, 75-79
Air quality monitoring stations 67, 68
Archaeology 3, 37, 59, 84, 88, 106, 110, 147
Archival material 11, 12, 15
Archival record 9, 11, 14, 15, 17-19, 21, 23, 25, 27, 29, 31
Artefacts 101, 102
Ashmolean Museum 17-20, 25, 28-30, 35
Automatic monitoring site 67
Average size of headstones 115

B

Backdrops, creeper-clad building 15
Bedding planes 21, 124, 125
Biblical quotations 109, 110
Biomonitoring 66, 78, 79
Biopitting 63
Blistering 14, 15, 62, 124, 129
Building ashlar 62, 63
Building exteriors 15, 34
Building façades 15, 17, 19, 21
Buildings 14, 16, 18-20, 23-25, 28-30, 64, 67, 81, 83, 84, 105, 136, 143
Burial grounds 84, 131, 134, 135, 141
Burials, non-headstone 91

C

Calibrated images 63, 147
Calibration, photographic 12
Camera-captured images 63
Canongate 129, 135, 136
Capillary 47, 49, 129
Capturing headstones 87
Cardinal readings 85, 136
Caveats 11, 65, 67, 77
Cemeteries 84, 87-89, 91, 92, 96-98, 102, 105, 134
   largest 97
Century dynamics 38
Century epitaphs 125

Century headstones 88
Century-scale 12
Century-scale evaluations 15
Century timescale 12
Changing Styles of Inscriptions on Headstones in English Urban Churchyards 134
Changing Tradition of Motifs to Epitaphs on Headstones 107
CIE Lab 61, 69
Clarendon Building 17, 21, 22, 28-30
Clarendon Building and Sheldonian Theatre 18, 20
Climbers 16, 18, 19, 81, 148
Climbing plants 15-20, 22, 23, 25-30, 63, 81
   environmental history of 18, 34
   evidence of 19, 81
Climbing vegetation 16, 19, 82
Close-up, free-style photographs 83
Close-up images 23, 33
Close-up of lichen colonisation 52, 53
Close-up photographic images 64
Close-up photographs 16
Close-up photography 8, 12
Colour chart 62, 69, 70, 147
Colour images 62
Colour photographs 69
Commemorating
   dateless headstone 91
   earliest headstone 91
   picture 91
Commemoration 87, 89, 92, 97, 99, 109, 131, 141
Convey photographic surveys 61
Corpse 118, 127, 128
Creeper 15, 16, 18-26, 28-30
Creeper growth 18, 23, 25, 28-30
Cross-spatial change 3
Cross-temporal change 3, 9, 11, 16, 83, 147
Crusts, black 47, 50
Cultural heritage 85, 131-133
Cultural record 107, 119, 125, 128

D

Damage headstone surfaces 58
Database, photographic 148
Death, motifs symbolising 117
Decay features 34, 62, 63
Decay mapping 61, 64
Digital cameras 4, 33, 35, 39, 61, 83, 91, 101, 102, 111, 136, 143
Digital databases 33, 35

U

Urban environments 3, 12, 147
Urban greening 15, 16, 81, 82
USB flash 106

V

Vandalism 58, 85, 101-103, 111, 119, 132, 139, 148
Vegetation 7, 13, 16, 52, 81, 82, 109, 143, 145, 147, 148
Vegetation cover 13, 16, 81, 143, 145, 147
Verbal epitaphs 107
   using 107
Vertical photography 13
   used close-range 13
Victorian headstones 125
Victorian period 57, 88, 115, 117, 125, 134, 139
ViewFinder 16, 18, 20, 27, 29, 31, 81
Virtual record 36
Visibility 23, 78

Visualising preservation issues 101, 103, 105, 107, 109, 111, 113, 115, 117, 119, 121, 123, 125, 127, 129

W

Walls 15, 19, 20, 24, 29, 61-64, 75, 76, 78, 105
   north-facing 63
Weathered headstones 47, 58, 103, 139
Weathered state 57, 58
Weathered stumps 139
Weathering 7, 15, 37, 39, 59, 81, 102, 103, 107, 110, 124, 125, 129, 132, 134, 139, 148
   cavernous 47, 49, 124, 125
   salt 124
Weathering and Stone Types 103
Weathering Limitations of Headstone Seriation 37
Wolvercote cemeteries in Oxford 89

Y

35-year-old restoration photographs 14
York's evidence of mortality 119

www.ingramcontent.com/pod-product-compliance
Lightning Source LLC
Chambersburg PA
CBHW041707210326

41598CB00007B/563